Mites of Greenhouses
Identification, Biology and Control

Other books on mites by Zhi-Qiang Zhang

1996 *The Eriophyoid Mites of China: An Illustrated Catalog and Identification Keys (Acari: Prostigmata: Eriophyoidea).* Co-author X.-Y. Hong

1997 *An Illustrated Guide to Mites of Agricultural Importance.* Co-author L.-R. Liang

1998 *Predatory Mites: Their Biology and Roles in Biological Control.* Co-authors J.-L. Xin & J.-Q. Lu

1999 *Tarsonemidae of China (Acari: Prostigmata): Annotated and Illustrated Catalogue and Bibliography.* Co-author J.-Z. Lin

2000 *Biology and Control of Bamboo Mites in Fujian.* Co-editor Y.-X. Zhang

2002 *Tarsonemidae of the World: Key to Genera, Geographical Distribution, Systematic Catalogue & Annotated Bibliograhy.* Co-author J.-Z. Lin

Mites of Greenhouses

Identification, Biology and Control

ZHI-QIANG ZHANG

Acarologist
Landcare Research, Auckland, New Zealand

Honorary Research Fellow
The Natural History Museum, London, UK

Adjunct Professor
School of Life Sciences, Fudan University, Shanghai, China

CABI Publishing

CABI is a trading name of CAB International

CABI Head Office
Nosworthy Way
Wallingford
Oxfordshire OX10 8DE
UK

Tel: +44 (0)1491 832111
Fax: +44 (0)1491 833508
Email: cabi@cabi.org
Web site: www.cabi.org

CABI North American Office
875 Massachusetts Avenue
7th Floor
Cambridge, MA 02139
USA

Tel: +1 617 395 4056
Fax: +1 617 354 6875
Email: cabi-nao@cabi.org

A catalogue record for this book is available from the British Library, London, UK.

Library of Congress Cataloging-in-Publication Data
Zhang, Zhi-Qiang, 1963-
Mites of greenhouses: identification, biology and control / Zhi-Qiang Zhang.
p. cm.
Includes bibliographical references and index.
ISBN 0-85199-590-X (alk. paper)
1. Plant mites. 2. Plant mites—Biological Control. 3. Greenhouse plants—Diseases and pests. 4. Mites as biological pest control agents. I. Title.

SB940.Z43.2003
635'.0496542—dc21

2003046117

ISBN-13: 978-0-85199-590-8
ISBN-10: 0-85199-590-X

First published 2003
Transferred to print on demand 2007

Printed and bound in the UK by CPI Antony Rowe, Eastbourne.

Table of Contents

Preface ix
Acknowledgements xi

Part I Introduction 1

1 Greenhouses, Plants & Mites 3
1.1 Introduction 3
1.2 Greenhouses 3
1.3 Plants grown in greenhouses 5
1.4 Mites in greenhouses 6
1.5 Recommended further reading 9
1.6 References cited 10

2 Introduction to Acari 11
2.1 General introduction 11
2.2 Morphology and structure 12
2.2.1 Division of body 12
2.2.2 Gnathosoma 13
2.2.3 Idiosoma 15
2.2.4 Legs 17
2.3 Classification 18
2.3.1 Higher classification 18
2.3.2 Order Prostigmata 23
2.3.3 Order Mesostigmata 27
2.3.4 Order Astigmata 29
2.4 Recommended further reading 29
2.5 References cited 30

3 Methods & Techniques 33
3.1 Collecting mites 33
3.1.1 Collecting mites from plants 33
3.1.2 Collecting mites on/in substrate 36
3.1.3 Things to note when collecting 36

3.2 Preserving mites for study .37
3.3 Preparing mites for microscopic study .37
3.3.1 Clearing/maceration of specimens .37
3.3.2 Temporary mounts .38
3.3.3 Permanent mounts .39
3.3.4 Recommended further reading .42
3.4 Rearing mites .42
3.5 References cited .42

Part II Pest Mites .45

4 Spider Mites .47
4.1 Introduction .47
4.2 Morphological characters .47
4.3 Life history and biology .48
4.4 Species important in greenhouses .50
4.4.1 *Tetranychus urticae* (Koch) .54
4.4.2 *Tetranychus cinnabarinus* (Boisduval) .61
4.4.3 Other spider mites .64
4.5 References cited .75

5 False Spider Mites .87
5.1 Introduction .87
5.2 Morphological characters .87
5.3 Life history and biology .88
5.4 Species important in greenhouses .90
5.4.1 *Brevipalpus obovatus* Dannadieu .90
5.4.2 *Brevipalpus phoenicis* (Geijskes) .93
5.4.3 Other species .96
5.5 References cited .97

6 Tarsonemid Mites .99
6.1 Introduction .99
6.2 Morphological characters .99
6.3 Life history and biology .100
6.4 Species important in greenhouses .101
6.4.1 *Polyphagotarsonemus latus* (Banks) .104
6.4.2 *Phytonemus pallidus* (Banks) .109
6.4.3 Other tarsonemid species .114
6.5 References cited .121

7 Eriophyoid Mites .127
7.1 Introduction .127
7.2 Morphological characters .127

7.3 Life history and biology ... 128
7.4 Species important in greenhouses ... 128
7.4.1 *Aculops lycopersici* (Massee) ... 129
7.4.2 *Epitrimerus alinae* Liro ... 133
7.4.3 Other eriophyoid mites ... 135
7.5 References cited ... 137

8 Acarid Mites ... 141
8.1 Introduction ... 141
8.2 Morphological characters ... 141
8.3 Life history and biology ... 142
8.4 Species important in greenhouses ... 142
8.4.1 *Rhizoglyphus robini* (Claparède) ... 147
8.4.2 *Rhizoglyphus echinopus* (Fumouze and Robin) ... 149
8.4.3 *Tyrophagus* species ... 151
8.4.4 *Mycetoglyphus fungivorus* (Oudemans) ... 158
8.5 References cited ... 158

9 Other Pest Mites ... 163
9.1 Introduction ... 163
9.2 Siteroptidae ... 163
9.3 Penthaleidae ... 165
9.4 Tydeidae ... 166
9.5 Tuckerellidae ... 166
9.6 Oribatida ... 167
9.7 References cited ... 168

Part III Beneficial Mites ... 169

10 Phytoseiid Mites ... 171
10.1 Introduction ... 171
10.2 Morphological characters ... 171
10.3 Life history and biology ... 172
10.4 Species important in greenhouses ... 173
10.4.1 *Phytoseiulus persimilis* Athias-Henriot ... 180
10.4.2 *Neoseiulus cucumeris* (Oudemans) ... 186
10.4.3 Other phytoseiid species ... 189
10.5 References cited ... 194

11 Laelapid Mites ... 203
11.1 Introduction ... 203
11.2 Morphological characters ... 203
11.3 Life history and biology ... 204
11.4 Species important in greenhouses ... 204

11.4.1 *Hypoaspis aculeifer* (Canestrini) 204
11.4.2 *Hypoaspis miles* (Berlese) 206
11.4.3 Other laelapid species 208
11. 5 References cited 208

12 Other Beneficial Mites 211
12.1 Introduction 211
12.2 Mesostigmata 211
12.2.1 Parasitidae 211
12.2.2 Ascidae 212
12.3 Prostigmata 213
12.3.1 Stigmaeidae 213
12.3.2 Anystidae 214
12.3.3 Cunaxidae 214
12.3.4 Erythraeidae 215
12.3.5 Tydeidae 216
12.4 References cited 216

Part IV Accessories 219

13 Glossary 221

14 Appendix: Mite Information Sources 229
14.1 Introduction 229
14.2 Professional societies/organizations 229
14.3 Serial publications 231
14.4 Directories of acarologists 233
14.5 Internet resources 233
14.6 Acarological collections 234
14.7 Acarology courses 234
14.8 Suppliers of natural enemies for biological control 234

15 Index 235

Preface

Mites are among the most important arthropods in greenhouses, both as pests causing economic injury to greenhouse crops and natural enemies used in the biological control of pest insects and mites in greenhouses. Because of their minute size, mites are much less well known than insects. Although a great deal has been written on mites of greenhouses in widely dispersed scientific literature, there is a lack of a comprehensive treatment dedicated to greenhouse mites. The need for such a book on the identification, biology and control of greenhouse mites has been brought to my attention by students, entomologists, pest control workers, and growers, indirectly through correspondence and directly through interactions with them in training courses and collaborative research.

This book is based on a training manual prepared by the author for a 'Short Course on Mites of Greenhouses', which was organized by the author and took place in the then CAB International Institute of Entomology (in the Natural History Museum, London, UK) three times during 1997-1998 (from 1999, the institute was integrated with other CABI institutes into CABI Bioscience). The course was primarily designed to help students, entomologists, pest control workers, and growers to identify mites that commonly occur on greenhouse crops, although the biology and control of major mite pests were also covered. The focus of the course on identification was warranted because any successful management of pests starts with the correct diagnosis of pest damage and identification of pest species, which are the key to any information about the species. In this book, however, the distribution, damage, biology and control of mite pests and roles of mite predators in biological control are covered, in addition to provision of simple diagnosis and user-friendly keys.

The book is divided into three main parts. The first part is an introduction to the book and includes three chapters. Chapter 1 is a general introduction to greenhouses, crops and mites. This is followed in Chapter 2 by an introduction to general mite classification, morphology and biology.

Methods for collecting, preserving and preparing mites for study are covered in Chapter 3. The second part of the book deals with the identification, biology and control of pest mites, including six chapters on spider mites, false spider mites, tarsonemid mites, eriophyoid mites, acarid mites and other pest mites, respectively. The third part of the book covers the identification, biology and application of beneficial mites in biological control and includes three chapters on phytoseiid mites, laelapid mites and other predatory mites, respectively. At the end of the book, a glossary is included and an appendix is provided with information on acarological journals, societies, courses and websites. An index is also included.

Zhi-Qiang Zhang
Auckland
May 2003

Acknowledgements

This book could not have been completed without the support of many people, to whom I am very grateful.

Prof. V.K. Brown, Director of the then International Institute of Entomology (IIE), for her strong support in my research on mites at the Institute and her initial recognition of the need for a training course in greenhouse mites, without which the manual for the course and this book could never have been prepared.

My colleagues at IIE; in particular, I would like to mention Dr Gillian Watson for reading the manuscript of the manual and her professional and moral support, especially during the difficult period of re-structuring of the Institute.

Mr Don MacFarlane, my predecessor at IIE and co-lecturer for another short course on mites of agricultural importance run annually at IIE, for his help and for sharing his experience and the manual prepared for that course.

My colleagues in the Department of Entomology, The Natural History Museum, especially Dr Anne Baker for her professional support and friendship and Keepers of Entomology, Richard Lane and Dick Vane-Wright, for the use of collections and facilities in the Department.

Participants of the course for their interest and enthusiasm in the course and feedback on the manual, which is very helpful.

Mr Tim Hardwick, CABI Book Publisher, for his patience and continuing support, especially when the preparation of this book was interrupted during the restructuring of IIE and my move to a new job at Landcare Research, New Zealand.

Tracy Ehrlich (CABI Publishing) for proofreading the manuscript and editorial assistance.

My colleagues Dr Anne Baker, Dr Nick Martin (Crop and Food Research, Auckland), Dr Trevor Crosby and Rosa Henderson (Landcare Research), and Dr Qinghai Fan (Fujian Agricultural and Forestry University,

Fuzhou) for the critical reviews of various chapters of the manuscript and helpful comments.

Mr Desmond Helmore for some of the illustrations used in the book and Rosa Henderson for technical assistance.

Mr Jianzhen Lin (Fujian Academy of Agricultural Sciences, Fuzhou, China) and Dr Anna Skoracka (Adam Mickiewicz University, Poznań, Poland) for help in obtaining literature.

Foundation for Science, Research and Technology, New Zealand for funding the author through the Invertebrate Systematics Programme (led by Dr Trevor Crosby) during the last stage of the preparation of this book.

Last but not least, I would like to mention and acknowledge three colleagues, who share my interests in greenhouse mites: Dr John Sanderson (Cornell University, Ithaca, New York, USA), who first introduced me to the study of greenhouse mites and supervised my doctoral research on the integrated management of spider mites in greenhouses; Dr Robert Jacobson (then Horticultural Research International, UK) and Dr Nick Martin, whose support and collaboration has helped me to sustain my interest in greenhouse mites when my primary area of research has shifted from mite ecology and biocontrol in greenhouses to mite systematics in museums and collections.

Part I
Introduction

- Overview of world's greenhouses, greenhouse plants and associated mites
- Introduction to mite morphology and classification
- Identification of mite orders and families in greenhouses
- Methods for collecting and studying mites

Greenhouses, Plants & Mites 1

1.1 Introduction

This book is about mites of greenhouses. Before we examine in detail the identification, biology and control of greenhouse mites, this chapter provides a brief introduction to greenhouses, plants grown in greenhouses and mites that attack greenhouse plants or are used in the biological control of pest mites and insects in greenhouses.

1.2 Greenhouses

Greenhouses are essentially light scaffolding covered by sheets of glass, fiberglass or plastic for maximum light transmission and heat retention. They are used in horticulture and floriculture to extend cropping seasons, to protect plants from adverse environmental conditions (e.g. extreme temperatures and storm), and to screen out plant pests and diseases (Hanan *et al.*, 1978). Their ability to avoid or prevent the invasion of pests and diseases, however, is limited, and thus there is the problem of plant diseases and pests, which include mites – the subject of this book.

Greenhouses first appeared when glass was used for covering houses. Thus they are also known as glasshouses. Even today the word *glasshouse* is used instead of *greenhouse* in the UK, although many of the so-called *glasshouses* in the UK are now covered by plastic. In this book, the word *greenhouse* is used to include both glasshouses in the strict sense and plastic houses. High walk-in plastic tunnels are similar to plastic houses in that they are high enough for a person to perform cropping practices from inside (Nelson, 1985). However, low plastic tunnels (row-covers) used to provide only seasonal protection to plants are not considered greenhouses here, because they are generally too low (often not more than 1 m in height) to allow a person to walk within, and general cropping practices must be performed from outside.

Depending on needs, climate and available resources, two main types of greenhouses are used in crop production (Fig. 1.1).

In the first type of greenhouse, environmental conditions are controlled at optimal levels for maximal crop production throughout the year. These greenhouses, often covered with glass, are expensive to maintain and are often used to produce high-value cash crops. They are most common in temperate areas (e.g. N. Europe and N. America).

In tropical/subtropical and Mediterranean areas, the other type of greenhouse (often under plastic) with minimal climate control is more widely used in crop production. They are less expensive to build and maintain, but can enable plants to grow better than they do in the field without protection and produce an economic yield (Enoch, 1986; Castilla, 1994).

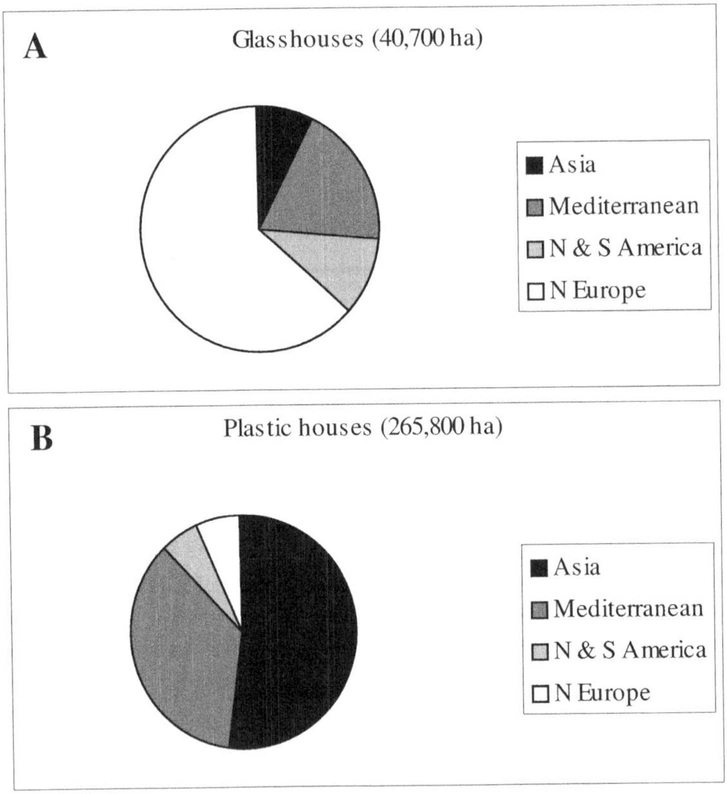

Fig. 1.1 Geographic distribution of glasshouses (A) and plastic houses (B) worldwide in 1995 (data from Wittwer and Castilla, 1995; high plastic tunnels are considered plastic houses here).

Greenhouses are becoming increasingly important in crop production in the world, especially in the production of vegetables, fruit crops and ornamental plants. The total surface area of greenhouses in the world nearly doubled during the 15 years from 1980 to 1995 (Table 1.1). By 1999, the greenhouse area in the world was estimated to be 307,000 ha, most of which (266,000 ha) was under plastic (Gullino *et al.*, 1999). In 1995, there were already more areas under plastic than under glass (Wittwer and Castilla, 1995; Fig. 1.1).

Table 1.1 Increase of greenhouse surface area from 1980 to 1995 worldwide.

Year	Total area (ha)	References
1980	150,000	Wittwer, 1981
1995	280,000	Bakker, 1995; Wittwer and Castilla, 1995

The geographic distributions of greenhouse areas showed significantly different patterns for plastic houses and glasshouses: most of the world's plastic houses were in Asia and the Mediterranean areas, whereas most glasshouses were in North Europe (Fig. 1.1).

1.3 Plants grown in greenhouses

A wide variety of economic plants are cultivated in greenhouses, including vegetable crops and ornamental plants (Table 1.2). Most crops are common ones cultivated in greenhouses throughout the world, and only some crops (e.g. grapes and tree fruits) are restricted to Asia (Japan), New Zealand and the Mediterranean.

The greenhouse industry initially developed in the UK, the USA and New Zealand before the Second World War. In the late 1950s and early 1960s, the greenhouse industry predominantly produced vegetables. Since then, there has been a gradual shift from vegetables to high-value ornamental crops in North America and Europe (Wittwer and Castilla, 1995). For example, the proportion of greenhouse area used for the production of ornamentals increased from 20% in the 1960s to 60% in the 1990s (Gullino *et al.*, 1999). In the USA, 95% of the greenhouse area is used for flowers, potted plants, ornamentals and bedding plants (Wittwer and Castilla, 1995). However, on the worldwide scale, vegetable crops still account for 65% of the total greenhouse area.

Table 1.2 Major economic plants commonly cultivated in greenhouses worldwide (modified from Wittwer and Castilla, 1995).

	Geographical distribution			
Crops	Asia	Mediterranean	N & S America	N Europe
Cucurbits	+	+	+	+
Strawberry	+	+	+	+
Solanaceous + green plants	+	+	+	+
Grapes + tree fruits*	+	+		
Lettuce, cabbage, celery, radish, asparagus	+	+	+	+
Flowers, ornamentals	+	+	+	+
Bedding + potted plants	+	+	+	+

* There is a small greenhouse grape industry in New Zealand, mainly for export to Japan.

1.4 Mites in greenhouses

Greenhouses are maintained under relatively stable environmental conditions for optimal growth of plants. Unfortunately, these conditions also favour the rapid growth of pests, especially small pests such as mites, which develop rapidly and can increase their population quickly over a short period before notice by growers. Culture methods commonly adopted in greenhouses also encourage pest development. For example, crops are often cultivated in monoculture in greenhouses, which facilitate the dispersal of pests among plants and rapid development of pest populations. The lack of natural enemies in the enclosed artificial environment and the rapid development of pesticide resistance in greenhouses are also important factors for the pest status of plant mites.

Mites, although much smaller in size than many insects, feature prominently in greenhouses, both as pests causing economic injury to crops (Table 1.3) and as predatory mites used in biological control of mite and insect pests (Table 1.4). This book discusses both pest mites (Part II) and beneficial mites (Part III), with emphasis on the biology and control of important pest species. Part I of this book includes this chapter, an introduction to mite morphology and taxonomy (Chapter 2) and reviews of methods and techniques for the study of greenhouse mites (Chapter 3).

Table 1.3 Common pest mites of importance in greenhouses.

Names	Affected crops	Importance
Tetranychus urticae Twospotted spider mite	Many vegetables (e.g. tomato, cucurbits, capsicum, French bean, strawberry, etc.) and ornamentals (rose, carnation, cyclamen, gerbera, etc.)	***
Tetranychus cinnabarinus Carmine spider mite	Mainly tomato and carnation	***
Polyphagotarsonemus latus Broad mite	Many vegetables and ornamentals	***
Phytonemus pallidus Cyclamen mite	Mainly cyclamen and strawberry	**
Aculops lycopersici Tomato russet mite	Tomato	**
Rhizoglyphus spp. Bulb mites	Mainly lily	**

Table 1.4 Common commercially available predatory mites of importance in biological control in greenhouses.

Names	Family	Pests attacked
Phytoseiulus persimilis	Phytoseiidae	Spider mites (*Tetranychus*)
Neoseiulus californicus	Phytoseiidae	Spider mites (*Tetranychus*)
Neoseiulus cucumeris	Phytoseiidae	Thrips, broad mite (*Polyphagotarsonemus latus*), cyclamen mite (*Phytonemus pallidus*)
Iphiseius degenerans	Phytoseiidae	Thrips, spider mites
Hypoaspis aculeifer	Laelapidae	Bulb mites (*Rhizoglyphus*), sciarid flies, fungus gnats, thrips
Hypoaspis miles	Laelapidae	Bulb mites (*Rhizoglyphus*), sciarid flies, fungus gnats, thrips

Spider mites (Tetranychidae) are among the most important pests on many greenhouse plants (Chapter 4). False spider mites (Tenuipalpidae), which are related to spider mites, sometimes also become pests on greenhouse plants (Chapter 5). Several species of tarsonemid mites (Chapter 6) and eriophyoid mites (Chapter 7) injure both vegetable crops and ornamental plants. Acarid mites of the genus *Rhizoglyphus* damages the bulbs of flowers and stored roots of many crops (Chapter 8). Other pest mites of minor economic importance are reviewed in Chapter 9.

The predatory mites of the family Phytoseiidae (Chapter 10) and Laelapidae (Chapter 11) are used in the biological control of mite pests and thrips on many crops in greenhouses (Table 1.4). Other beneficial mites of potential significance are discussed in Chapter 12.

Table 1.5 Common sense pest control methods in greenhouse crop production (modified from Dole and Wilkins, 1999).

Before production
Use pest-free plants and containers
Clean/sterilize benches and irrigation system
Remove weeds in and outside (within 3 to 9 m) of greenhouses
Remove fallen plant material, media, debris and extra plants
Use exclusion screens
During production
Use pest resistant species or cultivars
Inspect incoming plant material carefully; isolate new ones, if possible
Monitor pest populations regularly
Produce plants at the optimal growing conditions
Change clothes and boots before entering greenhouses
If a problem occurs or is likely to occur
Decide on the threshold of tolerance for each pest
Remove infested plants or plant parts promptly
Use beneficials when and where appropriate
Modify crop environment to discourage pest growth
Use effective chemicals when and where appropriate

In chapters on pest mites in Part II, importance is given to correct identification of pest species by their appearance, symptoms and morpho-

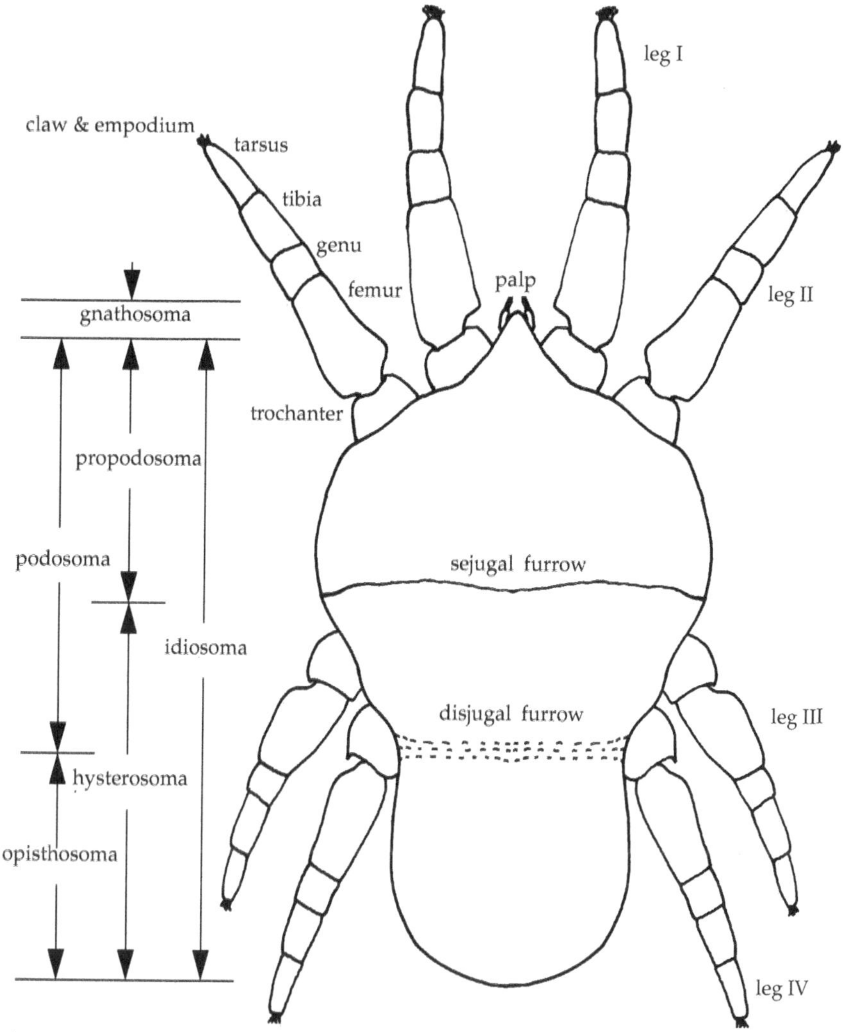

Fig. 2.1 Division of body in Acari as shown in a tenuipalpid male with small palps.

2.2.2 Gnathosoma

The gnathosoma is often anteriorly located, but sometimes hidden under the propodosoma. It is used mainly in handling food and feeding. The mouthparts comprise the chelicerae and the subcapitulum. The former lies above the oral opening, and the latter is formed by the fusion of palp-coxae, which enclose the feeding tube laterally and ventrally.

The chelicerae are used for cutting and piercing. They typically have three segments. Distal to the cheliceral base are a fixed digit dorsally and a basally hinged movable digit, which opposes the fixed digit (Fig. 2.2B). Cheliceral bases may be fused completely or partially; for example, they fuse to form the stylophore in spider mites (Fig. 2.2A). Most mites have a chelate-dentate chelicera which has teeth on opposing sides of the movable and fixed digits. Modifications of chelicerae occur in mites with different feeding habits. In phytophagous species such as the spider mites, the movable digit is modified into a stylet for piercing through the cell wall of the host plant (Fig. 2.2A). In some Mesostigmata, movable digits of the male chelicerae are modified for sperm transfer (Fig. 2.2B). The finger-like structure on the movable digits is called the spermatodactyl in Dermanyssina and spermatotreme in Parasitina in the Mesostigmata.

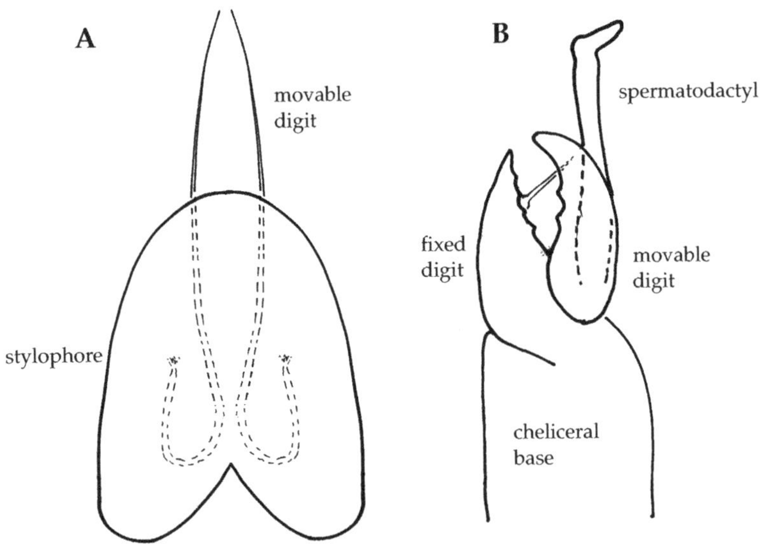

Fig. 2.2 Chelicerae. A, Tetranychidae, showing capsule-like stylophore and recurved, needle-like movable digit; B, Phytoseiidae (male), showing spermatodactyl.

The palps are used to feel and handle the food. They have a maximum of six free segments (Fig. 2.3): trochanter, femur, genu, tibia, tarsus and apotele (Fig. 2.3B). Losses (or fusions) and structural modification of palpal segments occur in various groups of mites and these have been used in classification. In some mites, the palptarsus is displaced from the the usual distal location to the ventral side of the tibia and forms a 'thumb-claw complex' together with a claw-like seta at the distal end of the tibia (Fig. 2.3A). This structure is used in holding food items. The palptarsi usually have sensory setae at the tip for detecting and/or tasting the food before or during feeding.

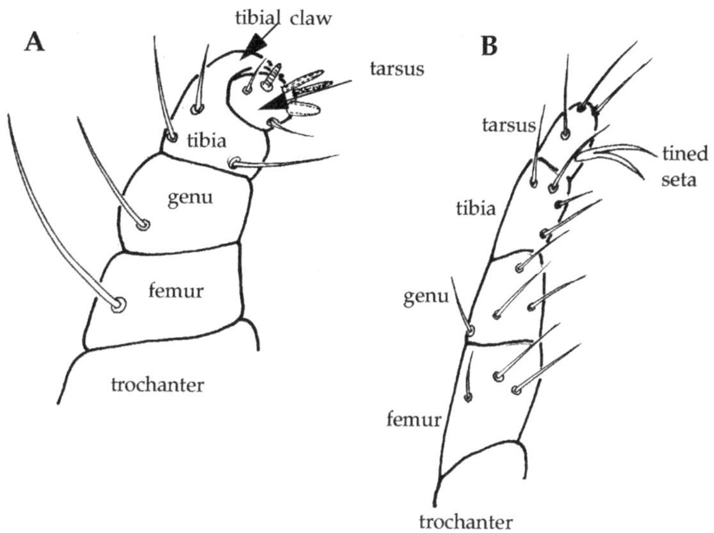

Fig. 2.3 Palps. A, Tetranychidae, showing 'thumb-claw complex'; B, Phytoseiidae, showing apotele (tined seta) on the palptarsus.

2.2.3 Idiosoma

The idiosoma is often ovoid or sac-like (Figs 2.1, 2.6-2.8) and occasionally worm-like as in the Eriophyoidea (Fig. 2.10A). It bears the legs and contains the organs for digestion, excretion and reproduction.

The cuticle of the idiosoma is often striate but may be covered with shields or plates (also known as scutum, scutellum, notogaster in various groups; Fig. 2.4), which vary among different groups of mites. The size, shape and degree of sclerotization of the shields have been used in classification. The shields sometimes can be very weakly sclerotized and the only difference between the shield and the surrounding cuticle may be just the absence of striae. The number and pattern of distribution of setae (chaeotrotaxy) on the surface of the idiosoma is important in taxonomy and has been used in classification in many groups. The structure of setae can vary greatly (Fig. 2.5) and is useful for classification. The relative length of setae and distance between setal bases are of importance at the species level in some families.

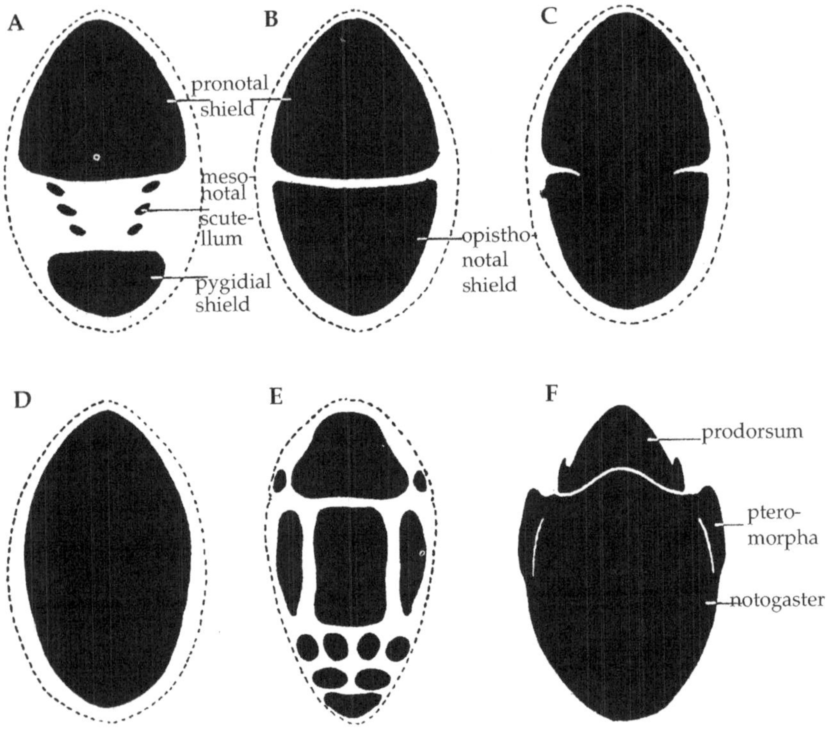

Fig. 2.4 Examples of dorsal sclerotization (denoted by black areas) in mites. **A**, protonymph; **B-D**, adults of a parasitid mite (Mesostigmata); **E**, adult of a stigmaeid mite (Prostigmata); **F**, adult of a member of the Palaeostomata (Oribatida) (reproduced from Evans, 1992 with publisher's permission).

Respiration may occur directly through the cuticle and/or via tracheae which open to the body surface by paired stigmata, often via tube-like or gutter-like peritremes. The tracheal system may open dorsolaterally (Fig. 2.7A), ventrolaterally, anteriorly near the base of chelicerae (Fig. 2.9.2A), or ventrally near the leg bases (Fig. 2.9.1A). The number and location of the stigmata and the shape and length of the peritremes have been used in higher classification.

Male and female genitalia usually open near the anterior end of the opisthosoma or between the leg coxae. When insemination is by spermatophores placed on the substrate by the male, there is little sexual dimorphism. When reproduction is by spermatophores transferred directly from the male to the female, males often vary greatly in size from females and genital openings vary in location.

The anus is usually located subterminally on the venter, rarely subterminally on the dorsum or terminally.

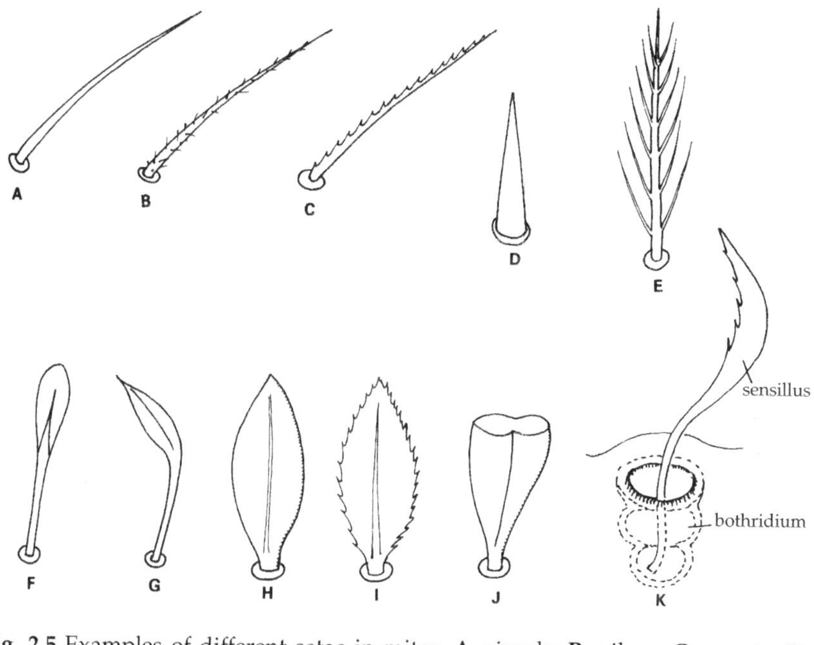

Fig. 2.5 Examples of different setae in mites. **A**, simple; **B**, pilose; **C**, serrate; **D**, spiniform; **E**, bipectinate; **F**, spatulate; **G**, falcate; **H**, lanceolate; **I**, lanceolate-serrate; **J**, cuneiform; **K**, bothridium and sensillus of an oribatid mite (reproduced from Evans, 1992 with publisher's permission).

2.2.4 Legs

Each leg consists of the coxa, trochanter, femur, genu, tibia, tarsus and apotele or ambulacrum (Fig. 2.1). The coxae are free and movable in the Parasitiformes, but are fused with the idiosomal venter in the Acariformes. Each femur may be divided completely or partially into two parts: basifemur and telofemur. Losses (or fusions) of leg segments occur in various groups of mites.

Legs have whirls of setae in dorsal, lateral and ventral positions. In addition to normal tactile setae, some mites (e.g. Acariformes) have specialized setae known as solenidion, eupathidium, famulus, microseta and trichobothria. Solenidia are sensillary in function and are usually found on the tarsi (designated by ω), tibiae (designated by φ) and genua (designated by σ). Eupathidia are specialized setae with a hollow interior. They are usually found on the tarsi and designated by ζ. Famuli are only found on the tarsi I and II of some mites. They are often very small in size and are designated by ε. Microsetae are found near the distal end of tibiae I and genua I and II in some mites and they are designated by κ. The kinds, number and pattern of distribution of setae (chaeototaxy) on leg segments (podomeres) are important for the classification in many groups of mites.

The ambulatory appendage at the distal end of the tarsus usually consists of a pair of lateral claws and an empodium, sometimes arising from an ambulacral stalk. Various modifications occur in different mite groups (Fig. 2.6) and these have been used in classification.

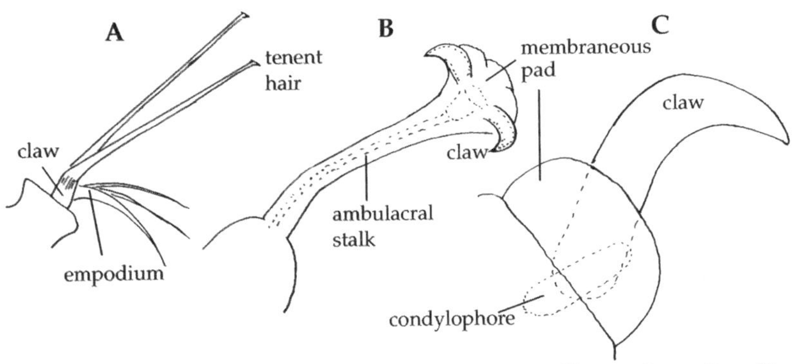

Fig. 2.6 Examples of ambulatory appendages in mites. **A**, *Tetranychus urticae* (Prostigmata: Tetranychidae); **B**, *Phytoseiulus persimilis* (Mesostigmata: Phytoseiidae); **C**, *Rhizoglyphus robini* (Astigmata: Acaridae).

2.3 Classification

2.3.1 Higher classification

Acari, also known as Acarina, has been considered as either an order or a subclass of the class Arachnida. Most acarologists now recognize Acari as a subclass, but the ordinal level classification is not settled (Krantz, 1978; Johnston, 1982; Lindquist, 1984; Evans, 1992; Walter and Proctor, 1999).

This book uses a traditional system adapted from Johnston (1982) and Evans (1992), recognizing three superorders and seven orders:

Superorder Opilioacariformes
  Order Opilioacarida (=Notostigmata)
Superorder Parasitiformes
  Order Holothyrida (=Tetrastigmata)
  Order Mesostigmata
  Order Ixodida (=Metastigmata)
Superorder Acariformes
  Order Prostigmata
  Order Astigmata
  Order Oribatida (=Cryptostigmata)

Some classifications have recognized superorders of this system as orders, and orders as suborders (e.g. Johnston, 1982). There are other classifications where Opiliocarida is grouped with Holothryrida, Mesostigmata (Gamasida) and Ixodida into Parasitiformes (e.g. Krantz, 1978; Lindquist, 1984). The Astigmata has recently been shown to be a subgroup within the Oribatida (Norton, 1998). Most members of the Endeostigmata, traditionally placed in the Prostigmata, are now considered more closely related to Oribatida-Astigmata (Sarcoptiformes) than Prostigmata. These views are now gaining support and acceptance (e.g. Walter and Proctor, 1999). However, for the ease of comparison with previous works, this book follows the traditional system above, which was used also by the author in a previous book on mites of agricultural importance (Zhang and Liang, 1997).

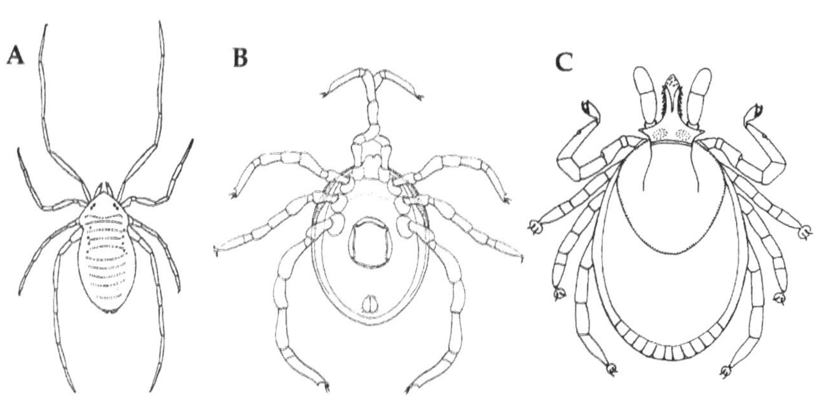

Fig. 2.7 Examples of **A**, Opilioacarida; **B**, Holothryrida; **C**, Ixodixa (after Evans, 1992).

The Opilioacarida is a small order of about 20 species distributed in Africa, Asia, Australia, Central America and Europe. They are large mites (1,500-2,300 µm) often found under stone and in litter. The legs are long and slender and are ornamented with bluish stripes and bands, with legs I and IV often longer than the body (Fig. 2.7A). The cuticle appears leathery. The body is constricted behind the podosoma. There are four pairs of dorsolateral stigmata. Very little is known about their biology and economic significance.

Mites of the Holothyrida are large (2,000-7,000 µm) predators inhabiting litter and soil in the southern hemisphere. The 30 or so described species are placed in nine genera and three families, and some 160-320 species are estimated to be present in the world (Walter and Proctor, 1999). The idiosoma is ovoid, covered dorsally by a single arched shield with a dense coat of short setae. Two pairs of lateral stigmata open in the intercoxal region (thus known in the past as Tetrastigmata). The female genital shield is in four distinct parts: a narrow pregenital shield anteriorly, a pair of laterogynal shields laterally and a large square postgenital shield posteriorly (Fig. 2.7B). The subcapitulum bears six pairs or more setae and a pair of stout corniculi. Distodorsum of tarsus I bears a cavity resembling Haller's organ in ticks. Little is known about the biology of Holothyrida and their economic significance.

The order Ixodida, commonly known as ticks, are large parasites (1,700-12,700 µm) of vertebrates. Some 880 species are known in the world, distributed in 12 genera and three families. A pair of stigmata is present near leg IV in nymphs and adults and because of this ticks were known as Metastigmata. Ticks are characterized by the hypostome armed ventrally and laterally with rows of recurved denticles (Fig. 2.7C) for holding fast on the hosts. Many species of ticks are vectors of diseases and of importance in veterinary medicine and human health.

The Oribatida are medium-sized (200-1,200 µm) mites and most common in soil and litter. They feed mostly on fungi and decaying plants, and are important for litter decomposition and soil formation. A few species spend part of or their whole lives on the aerial parts of plants. Some 11,000 described species of the world are placed in about 1,100 genera and 150 families. The oribatid mites have well sclerotized idiosoma (Fig. 2.8) and no obvious stigmata (hence were known as Cryptostigmata). The prodorsum bears up to six pairs of setae including a pair of sensilla arising from a deep conical base known as the bothridium (Fig. 2.5K). The chelicerae are chelate-dentate. The infracapitulum bears two to three pairs of adoral setae and four to seven infracapitular setae, of which one pair, the rutella, are greatly enlarged. Oribatids are not known as pests of economic plants.

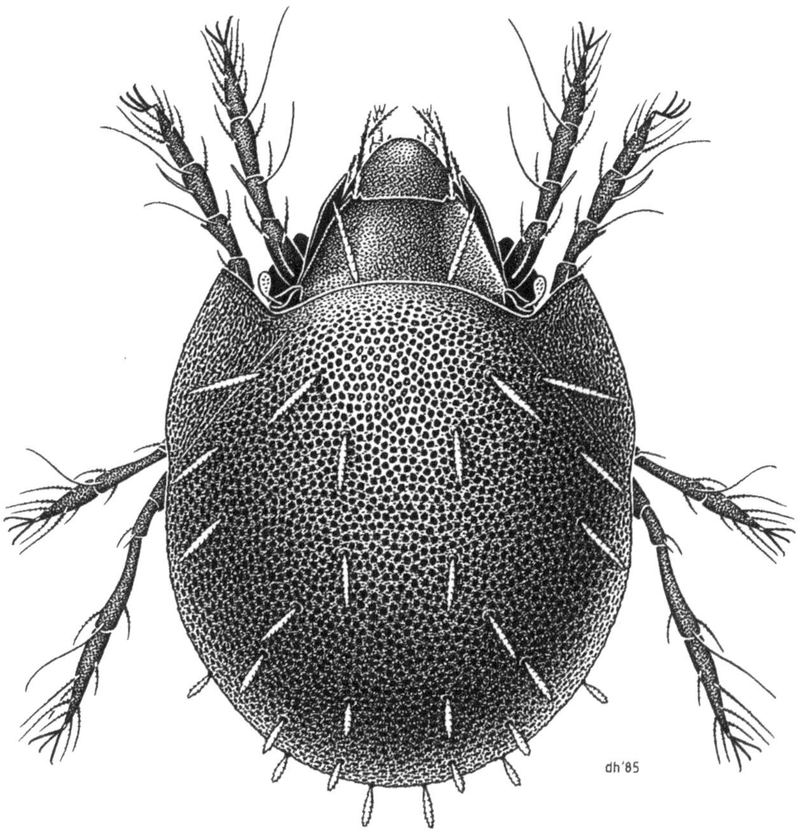

Fig. 2.8 Example of an oribatid mite: *Ramsayellus grandis* (illustrated by D. Helmore).

The four orders mentioned above are of no economic importance in greenhouses. Only mites of the orders Mesostigmata, Prostigmata and Astigmata are commonly found or released in greenhouses and are of economic importance in greenhouses. They can be separated using the illustrated key below (Fig. 2.9) and will be discussed in more detail.

Pest mites attacking greenhouse plants belong to the orders Prostigmata and Astigmata. Prostigmata also includes several families of predatory mites that attack pest species in greenhouses. The Phytoseiidae and other families of predatory mites of the order Mesostigmata also attack pest mites and other pests (insects and nematodes) in greenhouses.

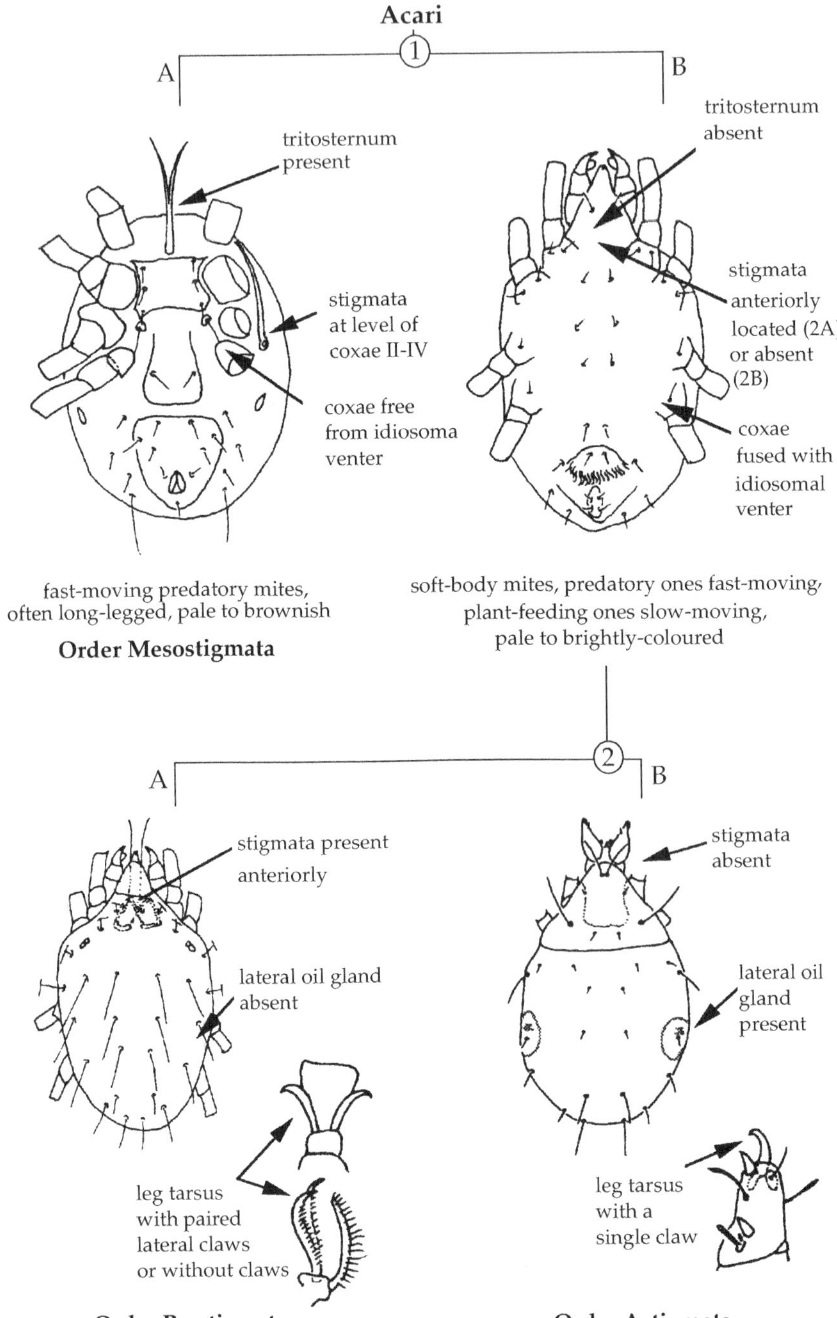

Fig. 2.9 Key to mite orders of importance in greenhouses (adults).

2.3.2 Order Prostigmata

This is the most diverse order with a great variation in body length (100-16,000 μm). They are found in a diverse range of habitats as predators, parasites, phytophages, and fungivores. Some 17,170 described species of the world are placed in 1,348 genera and 131 families (Walter and Proctor, 1999).

As the name suggests, the stigmata are anteriorly located on the anterior margin of the propodosoma or between the bases of the chelicerae (Fig. 2.9.2A). The palps may be modified by fusion or reduction of segments. The palptarsus may translocate to the basal part of the tibia to form a thumb-claw complex in some groups (e.g. spider mites; Fig. 2.3A). The fixed digits of chelicerae may be lost and the movable digit may be modified into a stylet (Fig. 2.3A). The cheliceral bases may be fused into a stylophore, or fused to the subcapitulum to form a stylophore-capsule. The subcapitulum bears both adoral setae and subcapitular setae. The sejugal furrow may or may not be present. There are often shields or plates on the dorsal propodosoma and hysterosoma (Fig. 2.4E). A maximum of five eyes may be present on dorsal propodosoma. Trichobothria may be present on the idiosoma and legs.

The typical life cycle of this order consists of egg, prelarva, larva, protonymph, deutonymph, tritonymph and adult stages, but various kinds of modification are common in different families. Development is by anamorphosis. The hexopod larva bears six opisthosomal segments, designated as C, D, E, F, H, and PS (pseudoanal), each bearing a row of setae. The protonymph adds the adanal (AD) segment and the fourth pair of legs, the deutonymph adds the anal (AN) segment, the tritonymph adds the peranal (PA) segment, and the adult adds the functional genitalia.

Mating is by stalked spermatophores placed on substrate by males (e.g. Erythraeidae) or by direct insemination (e.g. Tetranychidae). Sexual dimorphism is minimal in mites with indirect insemination.

Feeding habits are extremely diverse in this order. Mites of the families Tetranychidae (Chapter 4), Tenuipalpidae (Chapter 5), Tarsonemidae (Chapter 6), Eriophyoidea (Chapter 7) and Siteroptidae (Chapter 9) contain species that are injurious to economic plants in greenhouses. Mites of the families Stigmaeidae, Cunaxidae and Erythraeidae (Chapter 12) contain predatory mites that attack pest species in greenhouses and have potential as natural enemies in biological control. These families may be separated using the illustrated key below (Figs 2.10 and 2.11) and will be discussed in great detail later in various chapters on each group.

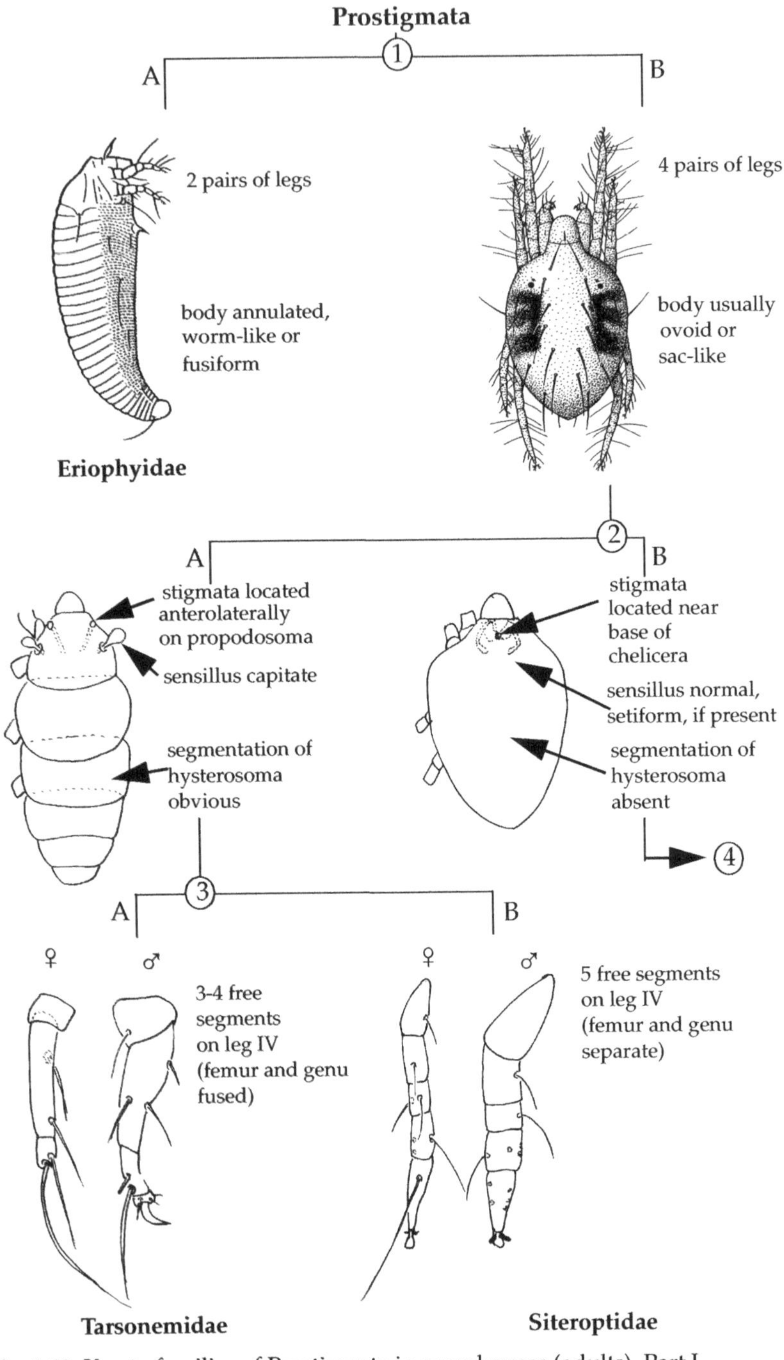

Fig. 2.10 Key to families of Prostigmata in greenhouses (adults). Part I.

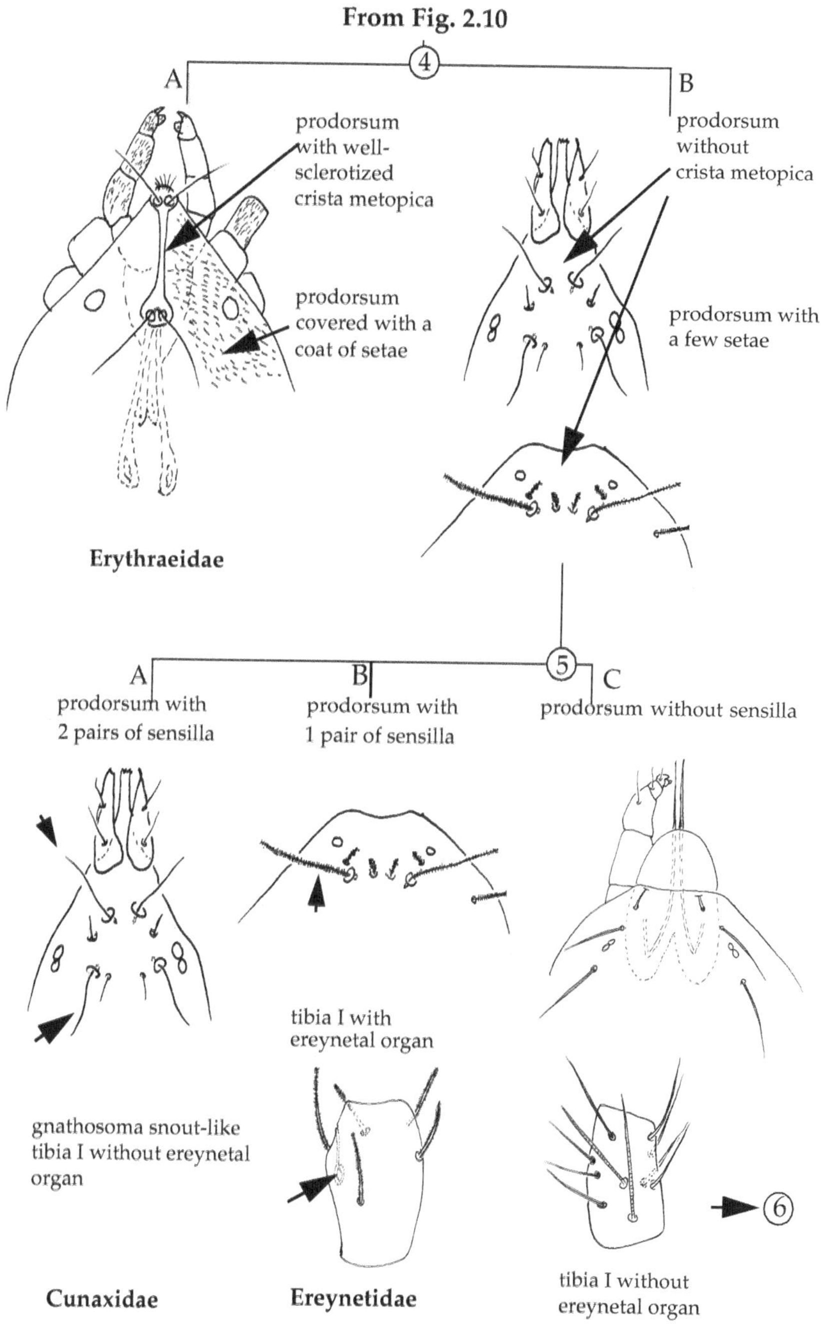

Fig. 2.11 Key to families of Prostigmata in greenhouses (adults). Part II.

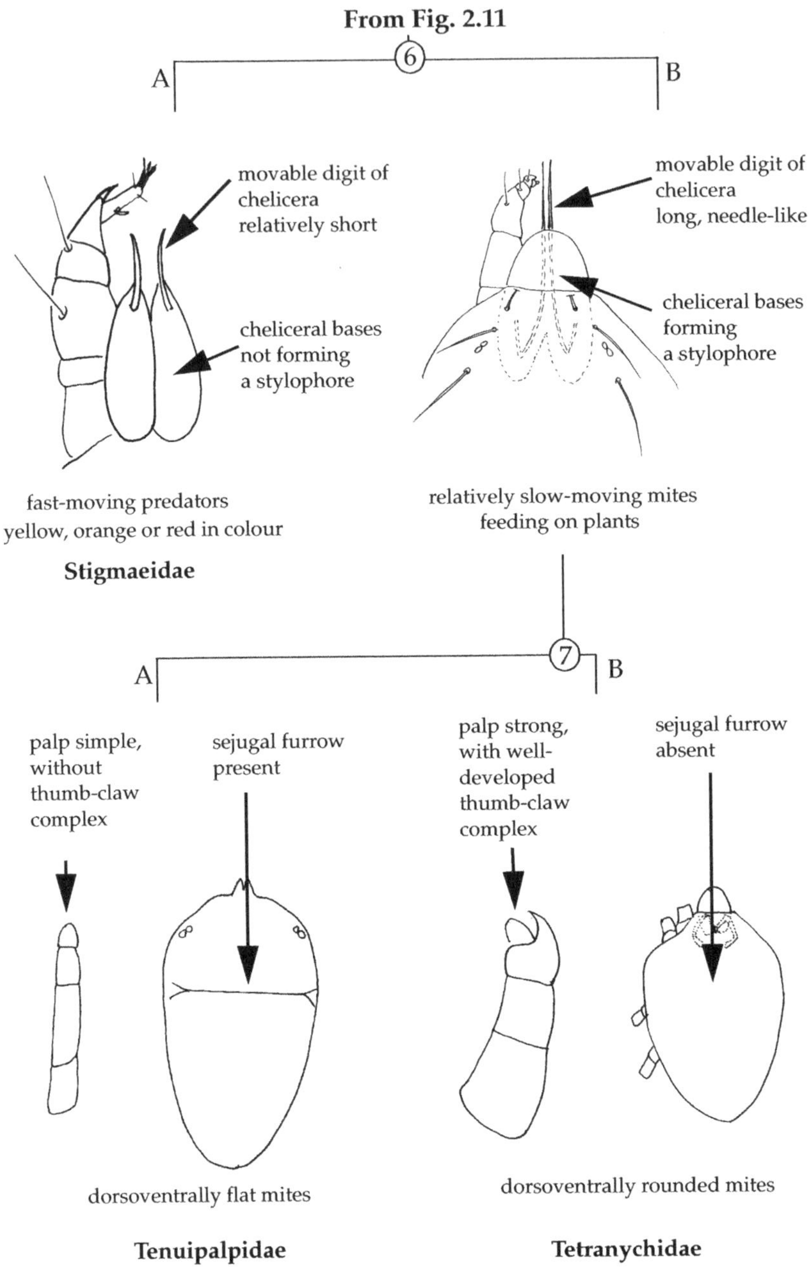

Fig. 2.12 Key to families of Prostigmata in greenhouses (adults). Part III.

2.3.3 Order Mesostigmata

These small to large (200-1,200 μm) mites are free-living predators in a variety of habitats, or parasites (or associates) of vertebrates and invertebrates. Many species are well sclerotized with brownish dorsal shields. Globally some 11,615 described species are placed in about 558 genera and 72 families (Walter and Proctor, 1999).

As the name suggests, the stigmata are located laterally or dorsolaterally in the region of coxae II-IV and a pair of associated peritremes is present in non-parasitic species. The idiosoma may be covered dorsally with one or two shields and ventrally by a sternal shield, genital and other shields. The female genital pore is covered by one to three sclerites. The male genital pore is sternal or presternal in position. The subcapitulum bears four pairs of ventral setae and corniculi. The palptarsus bears a two- or three-tined apotele on the inner, basal side. At the base of the gnathosoma, there is a tritosternum terminating anteriorly in a pair of laciniae or a single undivided lacinia. The laciniae lie above a denticulate deutosternal groove on the ventral subcapitulum. The chelicerae have dorsal and antiaxial lyrifissures. The fixed digit bears a seta, pilus dentilis.

Typical life cycle in the Mesostigmata consists of egg, larva, protonymph, deutonymph and adult stages. Dorsal sclerotization varies in different stages. The larva has a podonotal shield anteriorly, a small pygidial shield posteriorly and paired mesonotal scutellae in between (Fig. 2.4A). The nymphs and adult may have separate podonotal shield and opisthonotal shield (Fig. 2.4B), an incised schizodorsal shield (Fig. 2.4C) or a single holodorsal shield (Fig. 2.4D).

Reproduction can be by tocospermy, where sperm is transferred directly from the male genital orifice to that of the female, or by podospermy, where sperm is transferred by spermatodactyl (a pipette-like extension of the movable digit of chelicera; Fig 2.2B) to openings on coxae III of the female. Sex determination is by diplo-diploidy, haplo-diploidy or thelytoky.

Many species of Mesostigmata are free-living predators on plants or in soil. Four mite families, namely Phytoseiidae (Chapter 10), Laelapidae (Chapter 11), Ascidae and Parasitidae (Chapter 12) have been reported from greenhouses. The Phytoseiidae is by far the most important family of predatory mites that are widely used for biological control of spider mites and thrips. The Laelapidae contains species that may be useful for the control of mites attacking bulbs and fungus gnats. Some species of Parasitidae also show promise as biocontrol agents. These families may be separated using the illustrated key in Fig. 2.13.

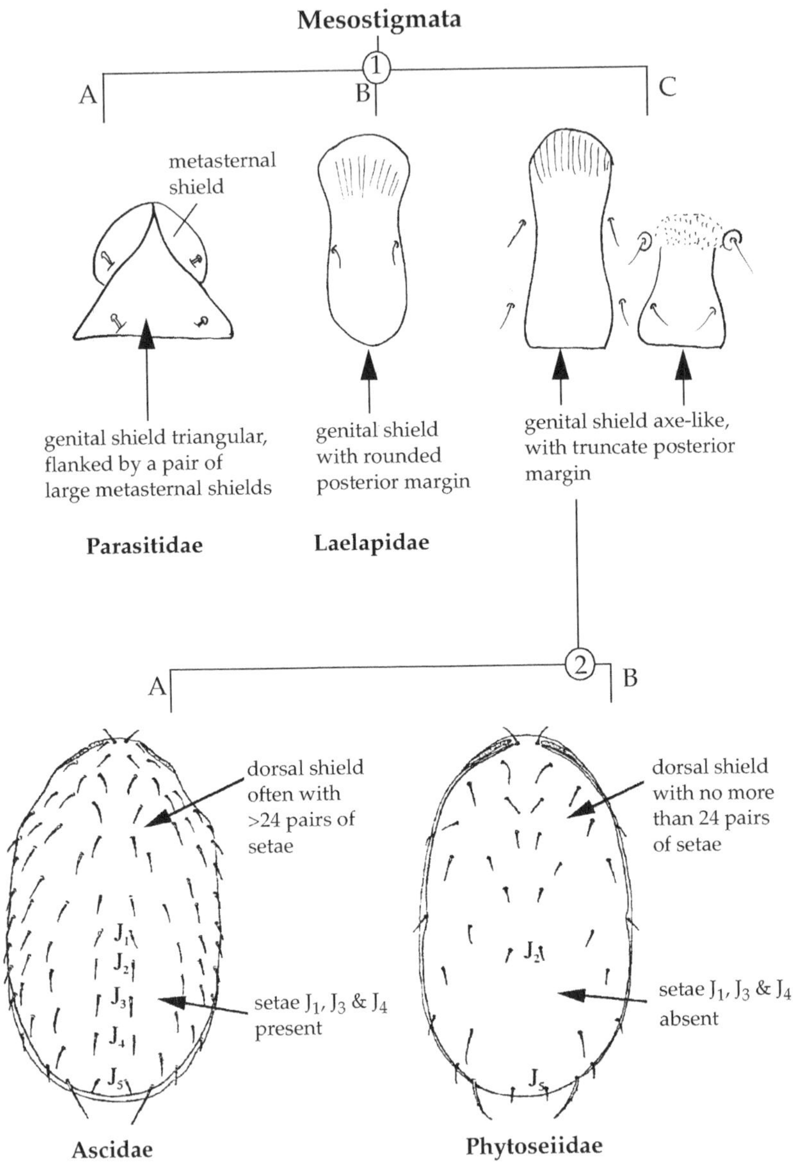

Fig. 2.13 Key to families of Mesostigmata in greenhouses (females).

2.3.4 Order Astigmata

These are usually weakly sclerotized, medium-sized (200-1,200 μm) mites found in a diverse range of habitats. Members of the suborder Acaridia are free-living or parasites, associated with insects or crustacea, and only rarely parasites of mammals, whereas those of the other suborder Psoroptidia are parasites of birds and mammals, rarely of insects, or free-living. Some 4,500 described species in the world are placed in 627 genera and 70 families (Walter and Proctor, 1999).

In this order, mites have no stigmata, as the name suggests. The subcapitulum has only one pair of setae. There is a single seta on the chelicera. The palps are reduced and have a maximum of two segments. The dorsal propodosoma bears a maximum of five setae and a Grandjean's organ near the base of trochanter IV; there are no trichobothria. The cuticle is often desclerotized, but a prodorsal shield may be present in some species (Fig. 2.9.2B). The sejugal furrow is often present. The opisthosoma bears a maximum of 12 pairs of setae on segments C to PS; segment F is without setae. Segment PA is absent. A pair of latero-abdominal glands are present in most free-living species. The legs have no true claws; there is often a claw-like empodium and a membranous ambulacrum (Fig. 2.6C; Fig. 2.9.2B).

The life cycle of this order is similar to that of the Prostigmata. A characterisitic feature of this order is a modification of the deutonymphs (commonly called hypopi). They are flattened, well sclerotized, without mouth and chelicerae, and are adapted for phoresy or resisting adverse conditions. There may be two morphs of males in some mites of this order. The homeomorphic males are similar to females but the heteromorphic males have enlarged legs III and/or IV.

Free-living mites of this order are of great importance in stored products. Only one family of this order is of importance in greenhouses. Several species of the family Acaridae (genera *Tyrophagus* and *Rhizoglyphus*) attack leaves and bulbs and corms of ornamental plants grown in greenhouses (Chapter 8).

2.4 Recommended further reading

Those who are interested in the phylogenetic position of the Acari and its relationships with related taxa should consult Weygoldt and Paulus (1979), Lingquist (1984) and Schultz (1990).

Evans (1992) provides a comprehensive review of the functional morphology of mites. Walter and Proctor (1999) emphasize the biology, ecology and behaviour of mites. Both these books lack systematic treatment of taxa below the ordinal level. As far as the systematic treatment at the family level and keys to families of mites are concerned, there is still no alterntiave to Krantz (1978, reprinted in 1986), although it is now somewhat outdated. However, the third edition of this is in preparation and should appear very soon.

Gerson and Smiley (1990) provide a concise summary of the biology and uses or potential role of mite families important in biological control. A key to families of mites of significance in biocontrol is also included. A Chinese edition of this book was published (Liang *et al.*, 1996) with an updated list of references. An updated and expanded edition of this book recently appeared as *Mites (Acari) for Pest Control* (Gerson *et al.*, 2003), with reviews of 34 families, a new glossary, and an updated list of about 1,600 publications.

Meyer (1981) reviews identification, biology and control of mites injurious to crops in Southern Africa.

Ochoa *et al.* (1991) provide an illustrated guide to phytophagous mites in Central America.

Ehara (1993) provides diagnostic features and life history data of many mites of agricultural importance in Japan, with excellent colour photographs of mites and symptoms.

Zhang and Liang (1997) give illustrated keys (in both English and Chinese) of many families and genera important in agriculture.

Students interested in the Prostigmata should read Kethley (1990), an excellent and important paper on this order with reviews of morphology, biology and keys to families in soil and litter. There is a lack of similar comprehensive treatments for the Mesostigmata and Astigmata. O'Connor's (1982) summary of familial classification of the Astigmata is worth reading for those interested in this order, but a key to families is lacking. A series of handbooks in Russian on the identification of soil-inhabiting mites are also useful on a more regional basis (Gilyarov and Krivolutsky, 1975; Gilyarov and Bregetova, 1977; Gilyarov, 1978).

2.5 References cited

Ehara, S. (ed.) (1993) *Plant Mites of Japan in Colors*. Zenkoku Noson Kyoiku Kyokai, Tokyo. 298 pp. [in Japanese].

Evans, G.O. (1992) *Principles of Acarology.* CAB International, Wallingford, UK. 563 pp.

Gerson, U. and Smiley, R.L. (1990) *Acarine Biocontrol Agents: An Illustrated Key and Manual.* Chapman and Hall, London. 174 pp.

Gerson, U., Smiley, R.L. and Ochoa, R. (2003) *Mites (Acari) for Pest Control.* Blackwell Science LTD, Cambridge. 539 pp.

Gilyarov, M.S. (ed.) (1978) *Handbook for the Identifcation of Soil-inhabiting Mites, Trombidiformes.* Zoological Institute of the Academy of Science, SSSR, Leningrad. 271 pp. [in Russian].

Gilyarov, M.S. and Bregetova, N.G. (eds) (1977) *Handbook for the Identifcation of Soil-inhabiting Mites, Mesostigmata.* Zoological Institute of the Academy of Science, SSSR, Leningrad. 717 pp. [in Russian].

Gilyarov, M.S. and Krivolutsky, D.A. (eds) (1975) *Handbook for the Identifcation of Soil-inhabiting Mites, Sarcoptiformes.* Zoological Institute of the Academy of Science, SSSR, Leningrad. 491 pp. [in Russian].

Halliday, R.B., O'Connor, B.M. and Baker, A.S. (1999) Global diversity of mites. In: Raven, P.H. and Williams, T. (eds) *Nature and Human Society.* National Academy Press, Washington DC, pp. 192-203.

Johnston, D.E. (1982) Acari. In: Parker, S.P. (ed.) *Synopsis and Classification of Living Organisms,* Vol. 2. McGraw-Hill, New York, pp. 111-169.

Kethley, J. (1990) Acarina: Prostigmata (Actinedida). In: Dindal, D.L. (ed.) *Soil Biology Guide.* John Wiley & Sons, New York, pp. 667-756.

Krantz, G.W. (1978) *A Manual of Acarology.* Oregon State University Book Store, Corvallis, Oregon. 509 pp.

Liang, L.-R., Zhong, J., Hu, C.-Y. and Dong, H.-Q. (translators) (1996) [*Gerson, U. and Smiley, R.L. (1990) Acarine Biocontrol Agents: An Illustrated Key and Manual. Chapman and Hall, London*]. Fudan University Press, Shanghai. 209 pp. [in Chinese, with foreword and updated references in English by Gerson and Smiley].

Lindquist, E.E. (1984) Current theories on the evolution of major groups of Acari and on their relationships with other groups of Arachnida, with consequent implications for their classification. In: Griffith, D.A. and Bowman, C.E. (eds) *Acarology VI,* Volume 1. John Wiley & Sons, New York, pp. 28-62.

Meyer, M.K.P. (Smith) (1981) Mites pests of crops in Southern Africa. *Science Bulletin Department of Agriculture and Fisheries Republic of South Africa* 397, 1-92.

Norton, R.A. (1998) Morphological evidence for the evolutionary origin of Astigmata (Acrai: Acariformes). *Experimental & Applied Acarology* 22, 559-594.

O'Connor, B.M. (1990) Astigama. In: Parker, S.P. (ed.) *Synopsis and Classification of Living Organisms,* Vol. 2. McGraw-Hill, New York, pp. 146-169.

Ochoa, R., Aguilar, H. and Vargas, C. (1991) Acaros fitofagos de America Central: Guia Ilustrada. CATIE, Costa Rica [in Spanish].

Schultz, J.W. (1990) Evolutionary morphology and phylogeny of Arachnida. *Cladistics* 6, 1-38.

Walter, D.E. and Proctor, H.C. (1999) *Mites - Ecology, Evolution and Behaviour.* University of New South Wales Press and CAB International, Sydney and Wallingford. 322 pp.

Weygoldt, P. and Paulus, H.F. (1979) Untersuchungen zur Morphologie, Taxonomie und Phylogenie der Chelicerata. 2 Cladogramme und die Entfaltung der Chelicerata. *Zeitschrift für Zoologische Systematik und Evolutionforschung* 17, 177-200.

Zhang, Z.-Q. and Liang, L.-R. (1997) *An Illustrated Guide to Mites of Agricultural Importance.* Tongji University Press, Shanghai. 228 pp. [in Chinese, with keys in Chinese and English].

Methods & Techniques 3

3.1 Collecting mites

Mites occur in greenhouses in two major habitats, on plants or in the substrate (soil or other growing media). The methods for collecting mites in these habitats are different. The collecting methods may also vary depending on the purpose: general observation/research and live/dead mites.

3.1.1 Collecting mites from plants

If plant material cannot be removed from the plants, many plant mites may be directly hand-picked from the plant with the aid of a fine hair brush (Fig. 3.1D), and with the aid of a hand lens if they are too small to be seen with the naked eye. A little water on the tip of the brush may help to hold the mites during transfer. Mites may be transferred to a rearing unit for further laboratory studies or into a vial with some kind of preservative (e.g. 70-80% alcohol) for later study.

If some plant material can be removed from the plants, a very simple way of collecting mites is to pick leaves or other plant parts, put them in a paper or plastic bag, and bring the sample to the laboratory for examination/collection under a microscope. During transport, bags with leaves may be stored in a thermal box with ice to reduce mite movement, desiccation, and predation by natural enemies. This method is widely used in both surveys and research. Mites can be examined or counted under a microscope and picked off with a hair brush for preservation in a suitable preservative or mounted directly in a mounting medium on a glass slide for microscopic examination (see Section 3.3 for details). Mites can also be collected live and used in starting a colony in the laboratory, or in a greenhouse for experimental studies.

For rapid-moving mites of relatively large size living on leaf and ground surfaces, a small aspirator may be used. It can be a simple one made of a large vial capped by a rubber bung with two tubes, one leading to a sucking source (a rubber bulb or the mouth of collector) and the other for pointing to the mites (Fig. 3.3). A gauze may be fixed to the end of the sucking tube inside the vial to prevent mites being sucked back out.

Fig. 3.1 Some useful tools for collecting and preparing mites. A, pipette; B, pin mounted on a plastic stick; C, forceps; D, fine hair brush (size 000).

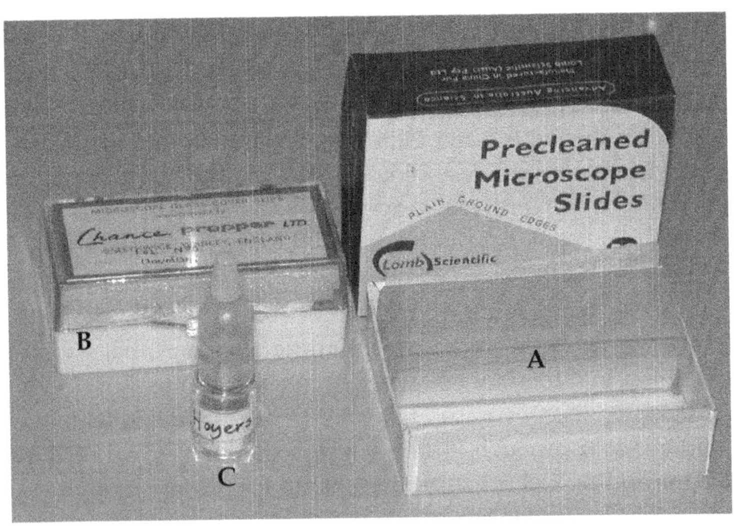

Fig. 3.2 Some slide-mounting necessities. A, glass slides; B, round or square cover slips (box shown here); C, drop bottle for mounting media.

Many mites may be too small to be seen against the background colour of the plant. Fortunately, many phytophagous mites and predatory mites living on leaf surfaces can be easily shaken onto a tray or a sheet of stiff paper when the plants are beaten with a stick or just with your hand if a stick is not available. Mites can be easily seen against the background and picked using a hair brush, if a black (or white depending on the colour of the mites which are to be collected) tray or sheet of stiff paper is used to hold the mites under the plant. This method is good for collecting large numbers of mites, but the drawback is that mites from different leaves or plants can get mixed up.

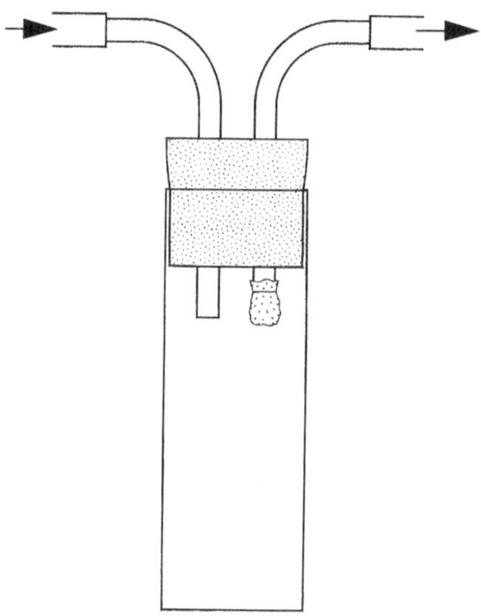

Fig. 3.3 An aspirator for collecting rapid mites and relatively large mites, with arrows indicating the direction of air flow (D. Helmore).

Rust mites and other free-living eriophyid mites can be collected by pouring a thin syrup of sorbitol (made of a 25% solution of isopropyl alcohol with a few iodine crystals) over the leaf or other plant part in a small container, which can be examined under a microscope. Gall mites can be collected by picking plant parts from the plant and putting them into small paper bags for later examination. Gall mites are best preserved as dried material in paper bags. These mummified mites can easily be recovered and made into good slides after many years.

Mites can also be collected by washing infested plants in hot water. A few droplets of detergent may be added to the water. When infested plant parts are shaken in water in a container, mites fall from the plants. Mites in the water may be separated by pouring the water through a sieve. Sieves of different mesh sizes can be used to separate mites of different sizes.

References & suggested further reading. Jeppson *et al*. (1975); Upton (1991); Amrine and Manson (1996); Perring *et al*. (1996).

3.1.2 Collecting mites on/in substrate

For collecting large numbers of mites from the surface of ground vegetation, a hand-operated vacuum apparatus can be used. The vacuum nets can be examined directly under a hand lens or dissecting microscope. Alternatively, the contents of the net may then be shaken onto a black (or white) tray and mites can be sorted and picked using a hair brush. If there is a lot of debris in the net, they may be washed into hot water and separated using the methods described above.

Large quantities of mites may be extracted from collected ground vegetation using a Berlese-Tullgren funnel (see Krantz, 1978 for an illustration of the apparatus). This method is extremely useful for getting qualitative data in faunal surveys and is very suitable for large collections of lower vegetation.

The Berlese-Tullgren funnel is also the most useful tool for extracting large quantities of mites living in soil and litter. Soil samples may be stored in paper or plastic bags during transport to the laboratory and spread over the mesh in the funnel for extraction.

References & suggested further reading. Krantz (1978); Upton (1991); McSorely and Walter (1991).

3.1.3 Things to note when collecting

When mites are collected for identification, it is important to collect a large sample of mites with different body sizes so that immatures and adults (males and females) will be represented in the sample. This is especially important for mites such as spider mites because males are required for identification at the species level for many species.

When collecting mites or any other insects, it is extremely important to write down data such as locality, date, collector and host(s). For the host, the scientific name, rather than the vernacular name, should be written on the label, if possible. Damage symptoms or feeding habits of the mites should also be noted if observed. For plant mites, any associated predatory mites should also be collected and the association recorded. For spider mites and many other mites, the colour of the mites should be recorded.

3.2 Preserving mites for study

Mites may be stored in small vials with 70-80% alcohol. Addition of 5% glycerol is recommended to prevent mites from drying out if the alcohol evaporates. Another preservative, the Oudemans' fluid, can also be used; it is a mixture of 87 parts of alcohol, five parts of glycerol and eight parts of acetic acid.

Vials used for storing should be small so that mites can be easily found later. Searching for small mites in a big jar can be difficult.

If possible, kill fresh mites by pouring onto them a small amount of boiling water so that mite appendages are fully extended. This will make later microscopic studies easier because structures on appendages can be seen most easily. Alternatively, fresh mites can be killed and fixed in a solution of methanol and acetic acid (two parts each plus one part distilled water) and this will also ensure that legs will be spread out. It is recommended that the dead mites be transferred from this solution to your regular preservative for storage within a week.

References & suggested further reading. Evans (1992); Saito and Osakabe (1992).

3.3 Preparing mites for microscopic study

3.3.1 Clearing/maceration of specimens

Mites are usually mounted on glass slides (Fig. 3.2A) for compound microscopic studies. Optical phase contrast and interference systems are sometimes necessary for examining fine structures. For very dark coloured specimens with a lot of body contents, it is necessary to clear the specimens before mounting them on slides. A common strong clearing agent is lactophenol, which is made from the following ingredients added in sequence:

Lactic acid	50 parts
Phenol crystal	25 parts
Distilled water	25 parts

Phenol is very caustic, so please be careful when mixing the chemicals. For not very engorged mites, lactic acid alone can be a very good clearing agent. An aqueous solution of 50-95% lactic acid works for most plant mites. Most mite specimens can be cleared in lactophenol or lactic acid for a week or so at room temperature. Smaller and soft-bodied mites require less time. To facilitate maceration of big mites, it is advisable to puncture the body using a fine insect pin. Maceration can also be greatly accelerated by heating the specimen in the clearing agent on a hot plate. Specimens can be cleared very quickly this way (in hours or minutes depending on temperatures of the hot plate and the size and sclerotization of the mite) and care should be taken that the specimen is not over-cleared. It is advised to transfer macerated specimens to distilled water to rinse off clearing liquid before mounting on slides.

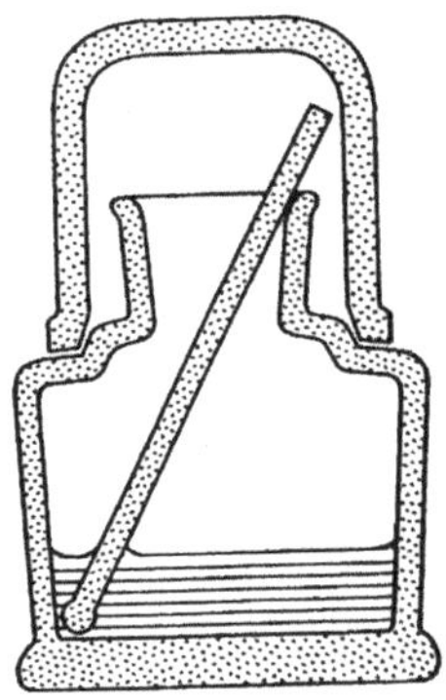

Fig. 3.4 A glass bottle for mounting media, showing a glass rod used for applying the mount to slides (D. Helmore).

3.3.2 Temporary mounts

Temporary mounts can be made using lactic acid. Mountants can be applied using a drop bottle (Fig. 3.2C) or a glass rod (Fig. 3.4). A droplet of the mountant should be placed in the centre of the glass slide. A mite is then added to the mountant using a minute insect pin (Fig. 3.1B). The mite can also be transferred to the mountant using a fine hair brush or an insect pin with a looped or spatulate tip. Once the mite is in the mountant, it should be oriented using an insect pin so that it is heading towards you with dorsal side upward. If you have extra specimens, some specimens may be mounted ventral side upward. For some mites, legs tend to curl under the idiosoma if they were not killed using methods described above and they need to be spread out carefully for easy microscopic study later. A cover slip (13 mm diameter is suitable for most mites) should then be lowered on to the specimen. Further adjustment of the orientation of the specimen may be made by gently moving the cover slip. The amount of mountant to put on the slide for each specimen is difficult to estimate at first. It depends on the size of the mite and the size of the cover slip. The aim is to let the cover slip only gently press the mite. With some practice, it can be easily done.

The mites can also be mounted on cavity slides for study. A square cover slip may be used to cover two-thirds of the cavity so that the other one-third allows a fine needle to be inserted to move the specimens for examination in different positions or from different angles. Examined specimens should be transferred to regular preservative for storage.

3.3.3 Permanent mounts

3.3.3.1 Hoyer's Medium

For more permanent mounts of specimens, Hoyer's medium is commonly used. The Hoyer's medium can be made by mixing the following ingredients in sequence:

Distilled water	25 ml
Gum arabic	15 g
Chloral hydrate	100 g
Glycerine	10 ml

It is important that the gum arabic used is from a crystalline source, but not a powdered form. The ingredients should be mixed at room temperature and the resulting liquid should be filtered using several layers of cheesecloth or bolting silk.

This medium is easy to use and also clears weakly sclerotized mites such as most plant mites. It has excellent optical properties, although it is not considered by some to be a permanent medium.

If lactic acid is used to clear specimens, then it is important to wash or soak the specimens in distilled water to remove the excess lactic acid and dissolved tissue before mounting the specimens in Hoyer's medium. This will reduce the chance that the slide will degrade quickly.

The Hoyer's medium is very toxic and care should be taken when using it. Contact with skin should be avoided. Wash hands thoroughly after using the medium or better wear a pair of gloves.

For some mites, correct orientation of the specimen is important. For example, males of spider mites of Tetrancyhinae often need to be mounted in a lateral position so that the male aedeagus can be in a lateral profile. This can be done by pushing/adjusting the cover slip when the mountant is still wet, which requires a lot of experience. A better and more reliable method for positional mounting is described by Henderson (2001):

1) Place a very small drop of Hoyer's medium on the slide and spread it out to a fairly thin layer.
2) Place a mite in the Hoyer's and with the aid of pins, position it lying on its side. There should be barely enough medium to coat the mite.
3) Before placing the cover glass, briefly dry the slide until the Hoyer's has set and the mite is firmly stuck in position. Drying can be in a drying oven at 40°C for up to 3 hours, or for longer periods at room temperature. Test the consistency of the medium on the slide with a pin: it should be impossible to spread the Hoyer's any longer, and the pin should make only a slight indentation in the surface. Do not worry that the mite appears shrivelled at this stage. On the other hand, do not dry until the medium is completely hard.
4) Place a fresh drop of Hoyer's medium on top of the set specimen, then gently lower a cover glass over them. As the fresh Hoyer's combines with the semi-dry medium, the mite(s) rehydrate while staying in their set lateral position.

Once properly mounted, it pays to use a marker pen to circle or mark the location of the mite on the slide if it is not in the centre. Mounted slides should be warmed at about 60°C until the specimens have fully cleared. Slides should be labelled as soon as possible with collection data including locality, date, collector and host(s).

3.3.3.2 Lactophenol Media

Lactophenol media, especially P.V.A., have also been used by some acarologists in place of Hoyer's medium, with varying results. A commonly used such medium is Heinze's P.V.A., which can be prepared using the following formulation:

Polyvinyl alcohol	10 g
Distilled water	40-60 ml
Lactic acid (85-92%)	35 ml
Phenol 1% aqueous solution	25 ml
Glycerol	10 ml
Chloral hydrate	100 g

Heinze's P.V.A. can be used in the same way as Hoyer's medium. It should be stored in a brown bottle.

3.3.3.3 Resin-Based Media

Resin-based media such as Canada Balsam and Euparal can make permanent slides but have limited applications in acarology. The disadvantages are the relatively poor optical properties and the difficult progress of full maceration and dehydration of specimens. In the past, some well-sclerotized mites and large mites were mounted in resin-based media, but Saito *et al.* (1993) recently showed that Canada Balsam works well for spider mites and other small mites.

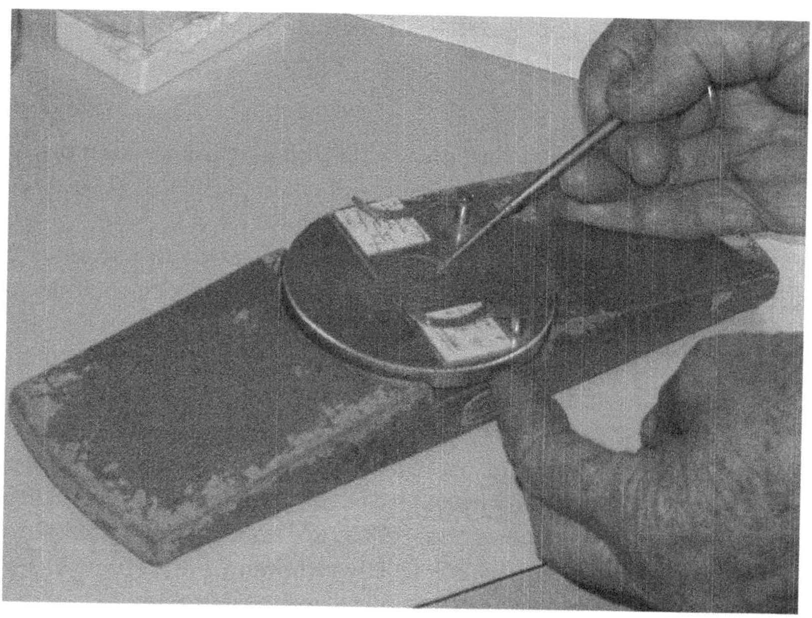

Fig. 3.5 Turntable for sealing cover slips on glass slides using a brush and insulating paint or other sealants.

3.3.3.4 Slide Drying And Ringing

Slides made from water-soluble media need to be fully dried. This can be done in a hot oven (40-50°C) for one or two weeks. If the air humidity of the collection room is not controlled to relatively low levels, then dried slides should be sealed around the cover slip using a kind of sealant. Glyceel ('Zut'), Euparal, and glyptal insulating paint have been recommended and used by many acarologists. Cover slips can be ringed using a small paint brush. The sealant should be evenly applied and the coat of the sealant should cover the complete circle of the cover slip. Several coats may be needed for good coverage. This can be most easily done using a turntable (Fig. 3.5), if the cover slip is round. Put the slide in the centre of the turntable, point the brush with ample sealant at the edge of the cover slip and turn the table. Repeat if necessary to get satisfactory results.

3.3.4 Recommended further reading

Singer (1967), Gutierrez (1985), Evans (1992) and Amrine and Manson (1996) for slide-mounting; Henderson (2001) for positional slide-mounting; Travis (1968), Tribe (1972) and Fain (1980) for slide-ringing; Upton (1993) for general reviews and comparision of different media.

3.4 Rearing mites

Rearing methods for mites vary greatly for different families of mites and for different purposes of rearing. These will not be further discussed here, but a guide to key information is included below.

Krantz (1978) provides a general review of the methods used for rearing various kinds of mites. Specific rearing methods and techniques are discussed and described: for spider mites by Helle and Overmeer (1985) and Lee *et al.* (1990); for tarsonemid mites by Liang (1980) and Xu *et al.* (1994); for eriophyoid mites by Oldfield and Perring (1996); for acarid mites by Ree and Lee (1997) and Okabe and O'Connor (2001); and for Phytoseiidae by Scriven and McMurtry (1971), Overmeer (1985), Scope and Pickford (1985), Piatkowski (1987), Brodeur and Cloutier (1992), Donia *et al.* (1995), Hadizadeh *et al.* (1997), Lee and Lo (1999), Heikal and Ali (2000), Rodriguez and Ramos (2000) and Shih (2001). Hughes (1976) describes methods for rearing acarid mites and predatory mites.

3.5 References cited

Amrine, J.W. Jr and Manson, D.C.M. (1996) Preparation, mounting and descriptive study of eriophyoid mites. In: Lindquist, E.E., Sabelis, M.W. and Bruin, J. (eds) *Eriophyoid Mites: Their Biology, Natural Enemies and Control. World Crop Pests*, Volume 6. Elsevier, Amsterdam, pp. 383-396.

Brodeur, J. and Cloutier, C. (1992) A modified leaf disk method for rearing predaceous mites (Acarina: Phytoseiidae). *Phytoprotection* 73(2), 69-72.

Donia, A.R.A., Helal, E.M., El-Hamid, M.M.A. and Zakzouk, E.A. (1995) Mass rearing and field evaluation of released predaceous mite, *Euseius scutalis* (Athias-Henriot) on the citrus whitefly, *Aleurotrachelus citri* (Priesner & Hosny). *Alexandria Journal of Agricultural Research* 40, 209-219.

Evans, G.O. (1992) *Principles of Acarology.* CAB International, Wallingford, UK, 563 pp.

Fain, A. (1980) A method of remounting of old preparations of acariens without raising or displacing the cover slip. *International Journal of Acarology* 6, 169-170.

Gutierrez, J. (1985) Mounting techniques. In: Helle, W. and Sabelis, M.W. (eds) *Spider Mites: Their Biology, Natural Enemies and Control.* Volume 1A. Elsevier, Amsterdam, pp. 351-353.

Hadizadeh, A., Daneshvar, H. and Kamali, K. (1997) Studies on mass production of *Phytoseiulus persimilis* Athias-Henriot in greenhouse conditions. *Applied Entomology and Phytopathology* 65, 22-24.

Heikal, I.H. and Ali, F.S. (2000) Mass rearing of the predaceous mite, *Phytoseiulus macropilis* (Banks) (Acari: Phytoseiidae). *Egyptian Journal of Agricultural Research* 78, 1477-1483.

Helle, W. and Overmeer, W.J.P (1985) Rearing techniques. In Helle, W. and Sabelis, M.W. (eds) *Spider Mites: Their Biology, Natural Enemies and Control.* Volume 1A. Elsevier, Amsterdam, pp. 331-335.

Henderson, R.C. (2001) Technique for positional slide-mounting of Acari. *Systematic & Applied Acarology Special Publications* 7, 1-4.

Hughes, A.M. (1976) *Mites of Stored Food and Houses.* Technical Bulletin of the Ministry of Agriculture, Fisheries and Food, No. 9. HMSO, London, 400 pp.

Jeppson, L.R., Baker, E.W. and Keifer, H.H. (1975) *Mites Injurious to Economic Plants.* University of California Press, Berkeley, California, 614 pp.

Krantz, G.W. (1978) *A Manual of Acarology.* Oregon State University Book Store, Corvallis, Oregon, 509 pp.

Lee, C.Y. and Lo, K.C. (1999) Mass rearing of Phytoseiulus persimilis Athias-Henriot (Acari: Phytoseiidae). *Chinese Journal of Entomology Special Publication* 12, 151-159.

Lee, W.T., Ho, C.C. and Lo, K.C. (1990) Mass production of phytoseiids: I. Evaluation on eight host plants for the mass-rearing of *Tetranychus urticae* Koch and *T. kanzawai* Kishida (Acarina: Tetranychidae). *Journal of Agricultural Research of China* 39(2), 121-132.

Liang, W.-J. (1980) Rearing method for rice-sheath-rot fungus free tarsonemid mite (*Steneotarsonemus spinki* Smiley). *Plant Protection Bulletin* (Taiwan) 22, 23-29.

McSorely, R. and Walter, D.E. (1991) Comparison of soil extraction methods for nematodes and microarthropods. *Agriculture, Ecosystem & Environment* 34, 201-207.

Okabe, K. and O'Connor, B.M. (2001) A method for both mass and individual rearing of fungivorous astigmatid mites (Acari). *Experimental and Applied Acarology* 25, 493-504.

Oldfield, G.N. and Perring, T.M. (1996) Rearing techniques. In: Lindquist, E.E., Sabelis, M.W. and Bruin, J. (eds) *Eriophyoid Mites: Their Biology, Natural Enemies and Control. World Crop Pests,* Volume 6. Elsevier, Amsterdam, pp. 377-381.

Overmeer, W.P.J. (1985) Rearing and handling. In: Helle, W. and Sabelis, M.W. (eds) *Spider Mites: Their Biology, Natural Enemies and Control.* Volume 1B.

Elsevier, Amsterdam, pp. 161-169.

Perring, T.M., Farrar, C.A. and Oldfield, G.N. (1996) Sampling techniques. In: Lindquist, E.E., Sabelis, M.W. and Bruin, J. (eds) *Eriophyoid Mites: Their Biology, Natural Enemies and Control. World Crop Pests,* Volume 6. Elsevier, Amsterdam, pp. 367-376.

Piatkowski, J. (1987) Mass rearing of the predatory mite *Amblyseius mckenzei* Sch. et Pr. *Materialy Sesji Instytutu Ochrony Roslin* 27(2), 43-48.

Ree, H.I. and Lee, I.Y. (1997) Development of mass rearing technique of *Tyrophagus putrescentiae* (Acari: Acaridae) found in house dust. *Korean Journal of Parasitology* 35(3), 149-154.

Rodriguez, H. and Ramos, M. (2000) Evaluation of rearing methods for *Amblyseius largoensis* (Muma) (Acari: Phytoseiidae) on *Polyphagotarsonemus latus* (Banks) (Acari: Tarsonemidae). *Revista de Proteccion Vegetal* 15(2), 105-108.

Saito, Y. and Osakabe, M. (1992) A new fixation method for preparing mite specimens for optical and SEM microscope observations. *Applied Entomology and Zoology* 27(3), 427–436.

Saito, Y., Osakabe, M., Sakagami, Y. and Yasui, Y. (1993) A method for preparing permanent specimens of mites with Canada Balsam. *Applied Entomology and Zoology* 28(3), 593-597.

Scope, N.E.A. and Pickford, R. (1985) Mass production of natural enemies. In: Hussey, N.W. and Scope, N.E.A. (1985) *Biological Pest Control. The Glasshouse Experience.* Blandford Press, Dorset, pp. 197-209.

Scriven, G.T. and McMurtry, J.A. (1971) Quantitative production and processing of tetranychoid mites for large scale testing or predator production. *Journal of Economic Entomology* 64, 1255-1257.

Shih, C.I.T. (2001) Automatic mass-rearing of *Amblyseius womersleyi* (Acari: Phytoseiidae). *Experimental & Applied Acarology* 25, 425-440.

Singer, G. (1967) A comparison between different mounting techniques commonly employed in acarology. *Acarologia* 9, 475-484.

Travis, B.V. (1968) Glyptal, a useful slide ringing compound. *Journal of Medical Entomology* 5, 24.

Tribe, H.T. (1972) Sealing of lectophenol mounts. *Transactions of the British Mycological Society* 58, 341.

Upton, M.S. (1991) *Methods for Collecting, Preserving and Studying Insects and Related Forms.* Australian Entomological Society, Brisbane.

Upton, M.S. (1993) Aqueous gum-chloral slide mounting media: an historic review. *Bulletin of Entomological Research* 83, 267-274.

Xu, D.-J., Deng, D.-F., Xia, X.-Y., Li, Y. and Luo, Q.-H. (1994) Methods for rearing and observing *Polyphagotarsonemus latus* (Banks) in the laboratory. *Yangtze River Vegetable* 1994(2), 16-17 [in Chinese].

Part II
Pest Mites

Identification, biology and control of

- Spider mites
- False spider mites
- Tarsonemid mites
- Eriophyoid mites
- Acarid mites
- Other pest mites

Spider Mites 4

4.1 Introduction

Spider mites belong to the family Tetranychidae of the order Prostigmata. They are so named because many members of this family produce silk webbing on host plants.

Spider mites are without doubt the most important mites attacking plants. Most spider mite species are polyphagous. They occur on virtually every major food crop and ornamental plant. They include some of the most important pests on economic plants.

The Tetranychidae is a large family of worldwide distribution. Some 1,200 species of spider mites belonging to over 70 genera are known in the world and more are yet to be discovered, especially in the Southern Hemisphere. The family consists of two subfamilies: Bryobinae and Tetranychinae. Most pest species belong to the Tetranychinae.

References & suggested further reading. Pritchard and Baker (1955); Helle and Sabelis (1985a); Meyer, (1987); Bolland *et al.* (1998).

4.2 Morphological characters

Spider mites are soft-bodied, medium-sized mites (about 400 μm for an average adult female). They are often red, green, orange or yellow in colour when alive.

The gnathosoma has a capsule-like structure known as the stylophore, which is formed by the fusion of the cheliceral body (Fig. 2.2A). The movable digits of the chelicerae are very long, often whip-like and recurved proximally. They are well suited for piercing. A pair of stigmata is located near the base of the chelicerae, where the peritremes arise. The palps are five-segmented (Fig. 2.3A). The palpal tarsus and tibia (with a distal claw-like seta known as the tibial claw) often form a thumb-claw complex. The tarsus often has an enlarged distal eupathidium (spinneret) in the Tetranychinae and this is used to spin webbing in many species. The size and shape of the spinneret is of taxonomic significance.

The idiosoma is often covered with a striate cuticle (Fig. 4.1). The pattern of the striation and the shape/density of lobes distributed on the striae are useful diagnostic characters.

There are three or four pairs of normal setae in two rows ($v_{1\text{-}2}$, $sc_{1\text{-}2}$) and two pairs of eyes on the dorsal propodosoma (Fig. 4.1).

On the opisthosomal dorsum, there are five rows of setae: *c*, *d*, *e*, *f* and *h* (Fig. 4.1). The number, location, length and structure of dorsal setae are of taxonomic significance.

Female genital pores are transverse and are bordered anteriorly by a genital flap and laterally by characteristic cuticular folds.

The structures of the paired lateral claws and the medial empodium are of taxonomic importance. The claws may be claw-like or pad-like with tenent hairs, and so is the empodium. Claws may bear dorsal or ventral hairs.

The tarsi of legs I and II bear duplex setae (a long solenidion and a short normal tactile seta with their bases joined together; Fig. 4.1). The number of duplex setae and their positions are of taxonomic significance.

Wedge-shaped males are smaller than ovoid females and have a tapering opisthosoma. Males have a protrudable aedeagus, the shape of which is very important in species identification.

References & suggested further reading. Lindquist (1985); Meyer (1987); Baker and Tuttle (1994).

4.3 Life history and biology

The life cycle of a spider mite consists of the egg, larva, protonymph, deutonymph and adult stages, with the exception of some *Schizotetranychus* and *Eotetranychus* species, which may have one nymphal stage in males. There are often quiescent intervals between each active stage, during which moulting takes place. Development from egg to adult often takes one to two weeks or more, depending on mite species, temperature, host plants, humidity and other environmental factors. Males develop slightly faster than females and find, guard and fight for quiescent deutonymph females. The winner mates with the female as soon as it emerges. Unfertilized eggs produce only males, which are haploid. Fertilized eggs produce diploid females.

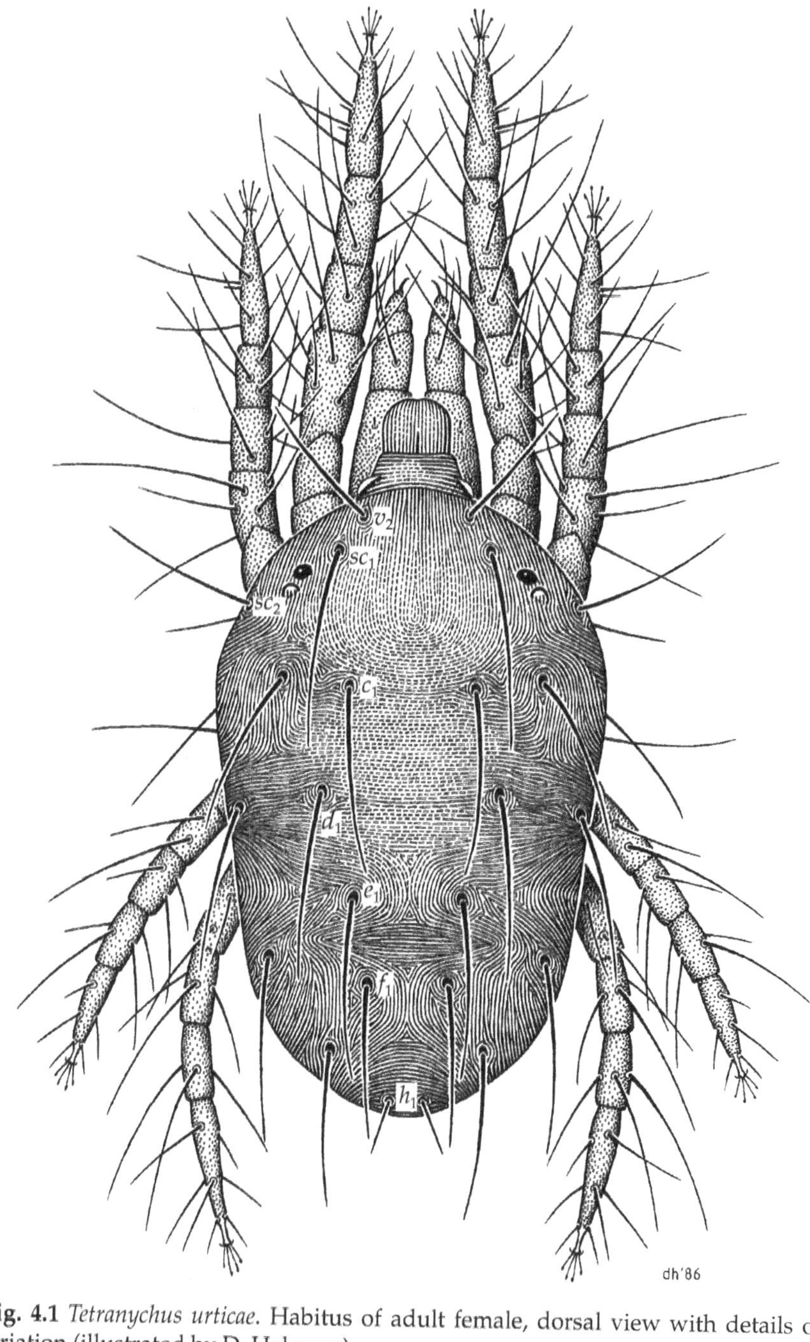

Fig. 4.1 *Tetranychus urticae.* Habitus of adult female, dorsal view with details of striation (illustrated by D. Helmore).

The life span of the adult female can be further divided into three periods: preoviposition, oviposition and postoviposition periods. The preoviposition period before the deposition of the first egg is usually short, lasting for a few days. The ovipositional period often lasts from ten to 40 days, during which the female can produce about ten eggs per day, with the maximum rate of reproduction occurring within a couple of days of the first egg. The postoviposition period is longer than the preoviposition period, but much shorter than the oviposition period.

Many spider mites prefer the under surface of leaves and feed by inserting their chelicerae into the leaves' parenchyma cells, the contents of which are then drawn into the body of the mite by a pharyngeal pump. Typical symptoms are small yellowish-white spots on the upper side of the leaf due to chlorophyll depletion, which develop into irregularly shaped white or greyish-coloured spots. The yellowing and bronzing of leaves may result. Necrosis may occur in young leaves and stems. Heavy infestation by some species may lead to leaf burning, defoliation, or even the death of the plant.

Spider mites may be spread from plant to plant by being carried in air currents. They may also be spread by the movement of infested plants or contaminated tools or clothing. They can also crawl to nearby plants, especially when the infestation is severe and leaves become dry.

Spider mites overwinter as females in the field. As day-length becomes shorter and temperatures become lower, adult females turn darker in colour and seek protected areas for hibernation. Often a chilling period is required before they reactivate and resume oviposition. In greenhouses, spider mites can reproduce throughout the year.

References & suggested further reading. Hussey and Huffaker (1976); Helle and Sabelis (1985a).

4.4 Species important in greenhouses

Most spider mites attack outdoor plants and a few of them regularly occur in greenhouses. Only *Tetranychus urticae* (Koch) and *Tetranychus cinnabarinus* (Boisduval) are widespread and frequently reach economic injury levels on many plants in greenhouses. A few species of the genus *Bryobia* are occasionally found in greenhouses and sometimes cause injury to plants. A couple of species of *Eotetranychus* also occasionally attack greenhouse plants, as does one species of *Panonychus* and another of *Oligonychus*. These are keyed in Figs 4.2 to 4.5.

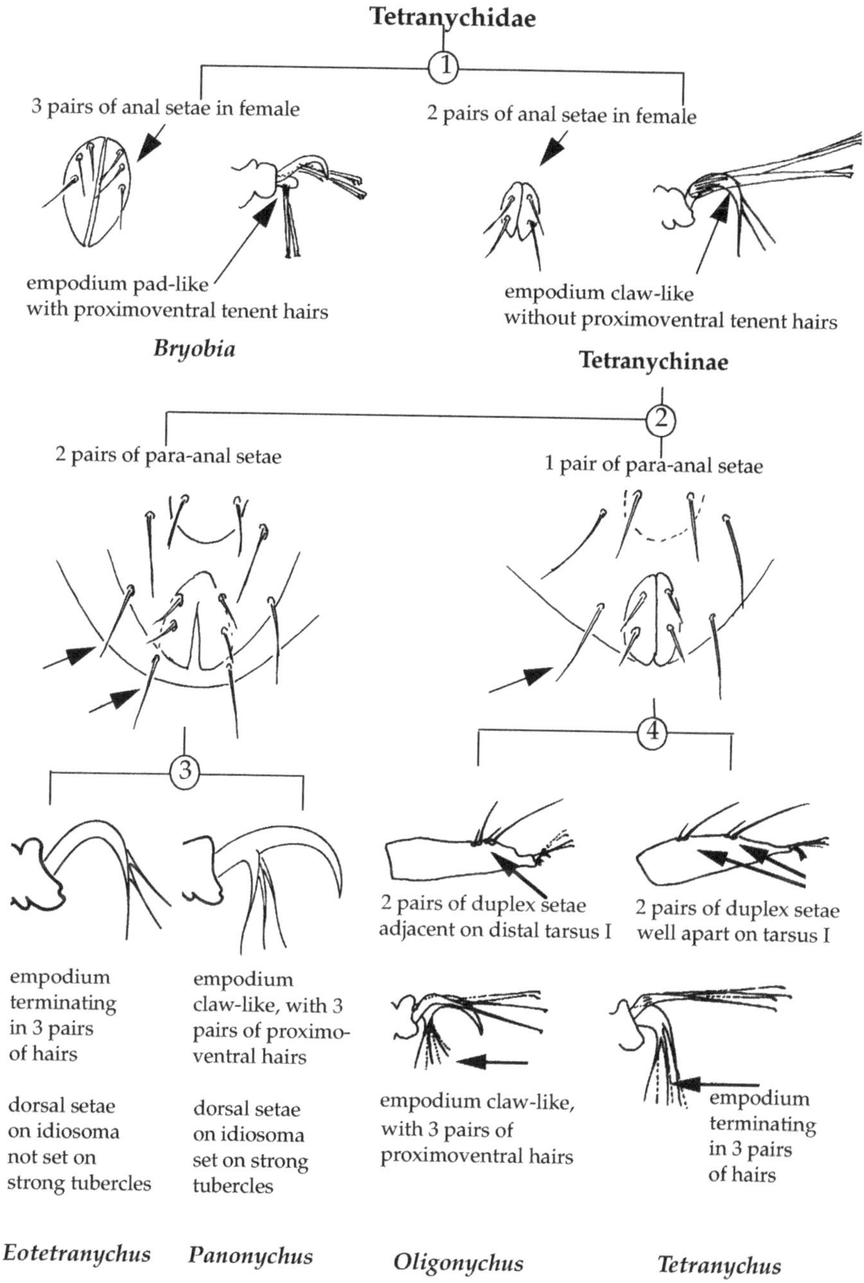

Fig. 4.2 Illustrated key to genera of greenhouse spider mites. For identification to species level, see Figs 4.3-4.5.

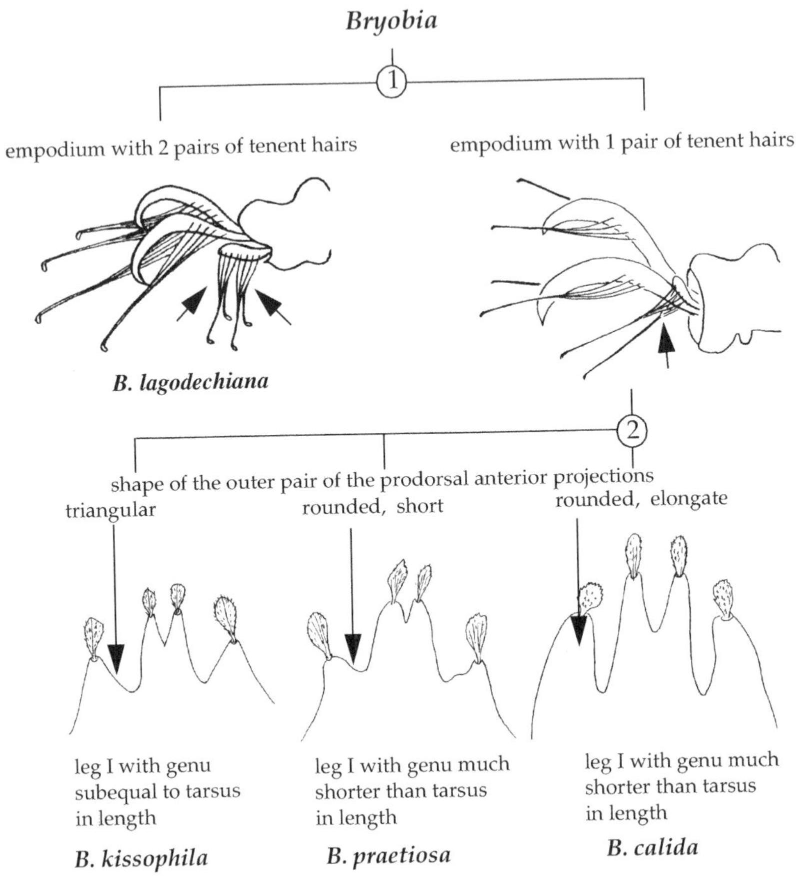

Fig. 4.3 Illustrated key to species of greenhouse *Bryobia*.

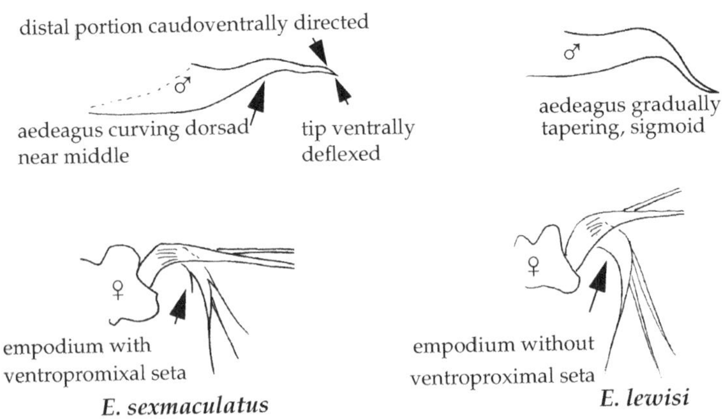

Fig. 4.4 Illustrated key to species of greenhouse *Eotetranychus*.

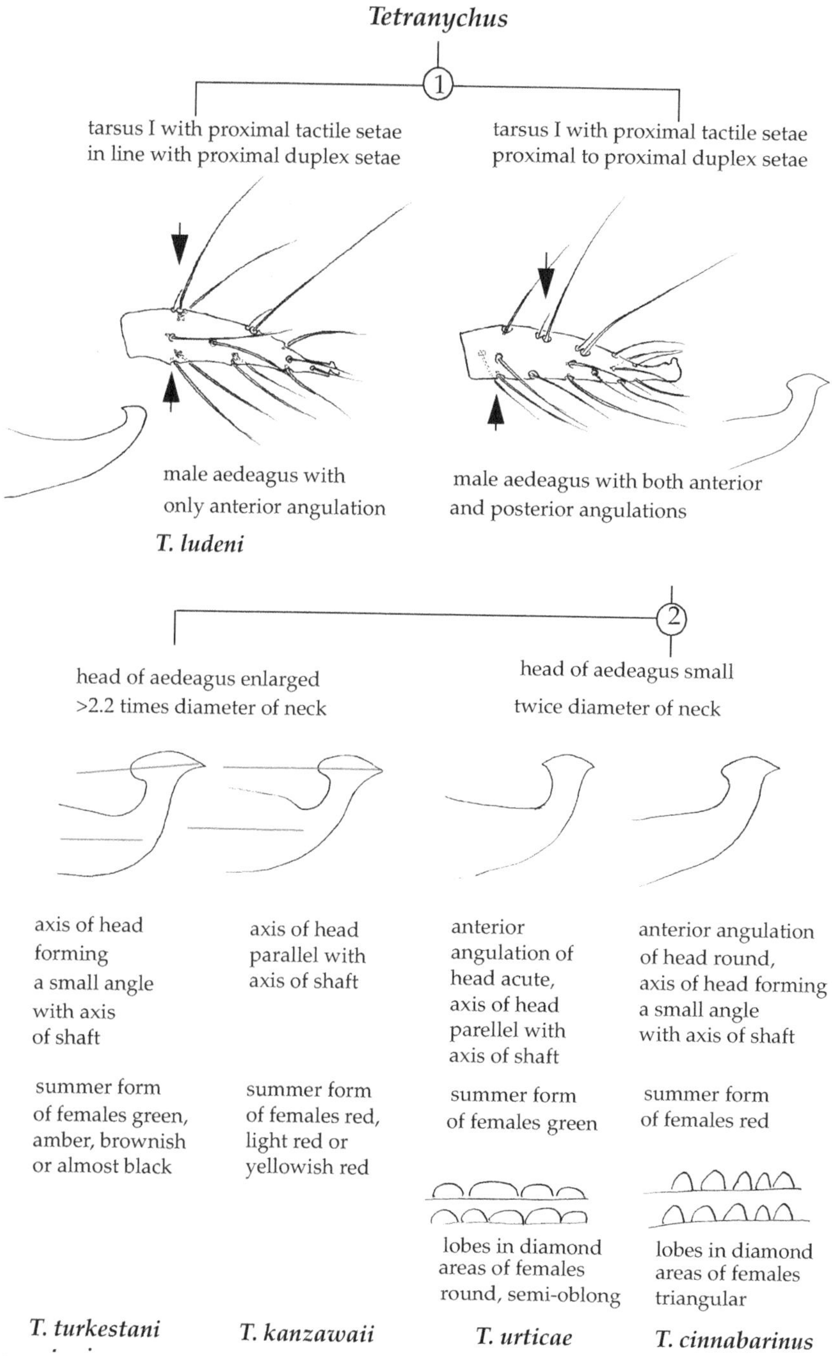

Fig. 4.5 Illustrated key to species of greenhouse *Tetranychus*.

4.4.1 *Tetranychus urticae* (Koch)

4.4.1.1 Common Name

The twospotted spider mite is the most common one. It is also known informally by many other names (e.g. the glasshouse spider mite, the yellow spider mite). Not very appropriately, it is often called 'red spider mite' or 'red spider' in some literature presumably because of the red/ orange colour of the overwintering form, or in reference to a species complex including *T. cinnabarinus*.

4.4.1.2 Distribution And Host Plants

This is a cosmopolitan species and common in greenhouses throughout the world. It is the most polyphagous species of spider mites and has been reported from over 150 host plant species of some economic value. It attacks over 300 plant species in greenhouses. A recent checklist includes some 1,200 host plant species in 70 genera for what these authors consider as *T. urticae*, which includes *T. cinnabarinus* (regarded by them as the red form of *T. urticae*).

References & suggested further reading. Jeppson *et al.* (1975); Bolland *et al.* (1998).

4.4.1.3 Appearance And Damage Symptoms

The eggs are often laid in clusters on the under surface of leaves. They are spherical in shape and translucent, pale in colour. As they develop, they become more yellowish and red eye spots inside the egg shell can be seen. Six-legged larvae are pale to yellowish when first hatched and become yellowish green after feeding. Eight-legged nymphs are yellowish green with dark spots, their body ovoid in shape with short legs. Adult females are about 400-500 μm and males are smaller with a tapered hysterosoma. The females (summer form) are yellowish to greenish in colour with two black spots on dorsolateral idiosoma (Fig. 4.6), but are darker in colour, often orange or red in the overwintering form. The colour of mites may vary depending on the host plant and other factors.

Twospotted spider mites often feed on cell chloroplasts on the under surface of the leaf. The upper surface of the leaf develops characteristic whitish or yellowish stippling, which may join and become brownish as mite feeding continues. As mites move around, their webbing can span leaves and stems. Heavy damage may cause leaves to dry and drop, and the plant may be covered with webbing and may die prematurely.

References & suggested further reading. Boudreaux (1956); Jeppson *et al.* (1975); Meyer (1981).

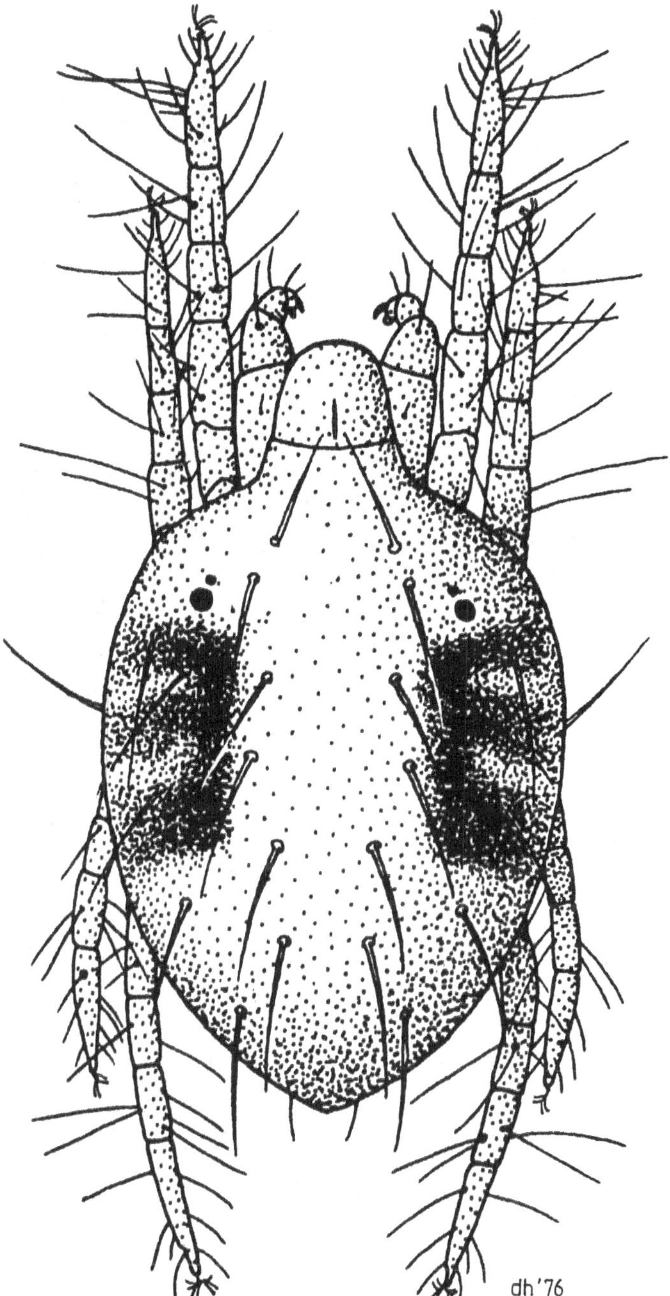

Fig. 4.6 *Tetranychus urticae.* Habitus of a young adult female, dorsal view, with details of body marking after feeding (illustrated by D. Helmore).

4.4.1.4 DIAGNOSTIC CHARACTERS

Adult female (summer form): there is a diamond pattern of striae between setae e_1 and f_1 on dorsal hysterosoma (Fig. 4.1). The lobes on striae in the diamond area are often rounded or semi-oblong and are often wider than tall (cf: often triangular in *T. cinnabarinus*; Fig. 4.5.2). The proximal tactile setae of tarsus I are proximal to the duplex setae as in Fig. 4.2 (cf: proximal tactile setae aligned with the proximal duplex setae in *T. ludeni*). Tibia I has only ten tactile setae (cf: often one to three solenidia in addition to ten tactile setae in *T. cinnabarinus*).

Adult male: empodium I is clawlike. The aedeagus is distinctive in having a small knob set at right angles to the neck; the anterior and posterior angulations are acute, small and equal (cf: the anterior angulation often slightly rounded in *T. cinnabarinus*; Fig. 4.5.2); the dorsum of the knob is rounded (cf: broadly angulated in *T. cinnabarinus*; Fig. 4.5.2).

References & suggested further reading. Boudreaux (1956); Brandenburg and Kennedy (1981); Meyer (1987); Kuang and Cheng (1990); Zhang and Jacobson (2000).

4.4.1.5 LIFE HISTORY AND BIOLOGY

Development occurs between 12 and 40°C. Developmental time from egg to adult decreases with increasing temperature (Fig. 4.7) and is less than a week at optimal temperatures for development (30-32°C). Under a diurnal temperature cycle of 15 to 28°C, developmental time is about 16 days. Males develop slightly faster than females.

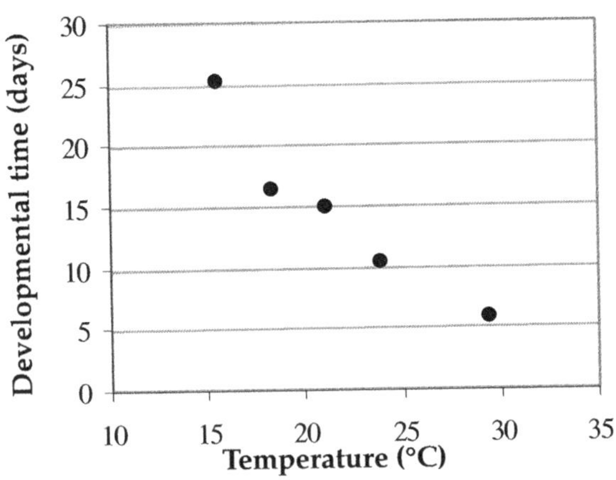

Fig. 4.7 *Tetranychus urticae*: developmental time from egg to adult in relation to temperature (drawn from data in Carey and Bradley, 1982).

Males are attracted to the sex pheromone from quiescent female deutonymphs. They guard their territory and fight fiercely against any other invading males. Mating occurs as soon as females emerge.

Females start to lay eggs within a couple of days of adulthood. The rates of oviposition and fecundity varies with food plant and temperature. An average female can lay over ten eggs per day and produce over 100 eggs during two weeks at about 25°C. The sex ratio is highly female biased, with a female to male ratio of about 3:1.

Tetranychus urticae disperses by active walking or by passive transport in the wind, on plants, on tools or on people.

Diapause is induced by short day length, lack of food supply and low temperature, and is normally terminated by a fixed period of chilling. Gravid females seek a protected niche at the end of summer. Diapausing adults are orange/red in colour.

References & suggested further reading. Laing (1969); Penman and Cone (1972); Shih *et al.* (1976); Carey and Bradley (1982); Rao *et al.* (1996); Bosse and Veerman (1996); Koveos and Veerman (1996).

4.4.1.6 Control and Management

Biological control

Biological control of *T. urticae* by phytoseiid mites is now widely used by the greenhouse industry in Europe, some parts of Asia and Africa, Australasia and North America. The most commonly used species are: *P. persimilis*, *P. micropilis*, *Neoseiulus californicus*, *N. fallacis*, *N. longispinosus* and *Galemdromus occidentalis*. They are often released repeatedly in biological control, but also inoculatively (such as in 'pest in first' methods, in which predators and spider mites were introduced together early in the season; see Chapter 10). The rates at which predators are released vary with the density of the spider mites, crop species, the temperature and other environmental conditions. The biocontrol companies supplying these predators often give useful instructions on how to use them on different crops. Biological control using phytoseiid predators on major greenhouse crops will be discussed in detail in Chapter 10 on the Phytoseiidae.

Predatory midges are also useful biocontrol agents against spider mites. A common species, *Feltiella acarisuga*, is commercially available for use in many countries. They are supplied and shipped by the biocontrol companies as cocoons on an inert substrate in a container. Release is best applied in the early morning or late evening, when it is cool and humid. Adults of *F. acarisuga* are mosquito-like but smaller in size. They mate within 24 hours of emergence and locate spider mite colonies, where they lay about 30 eggs over a five-day life span. Development occurs over a range of 15-25°C and is slowed down at 27°C. The eggs and larvae are killed at 30°C and above. The larvae hatch from eggs in a couple of days and feed on all stages of spider mites. They feed for four to six days and consume a total of about 150 spider mite eggs. They develop best when food is abundant, but can also pupate at a reduced size in times of food shortage. This predator can control spider mites on tomato and cucumber and the effects are enhanced by high humidities and a source of sugar droplets or honeydew. Repeated releases may be needed and rates will depend on prey density, crop and environmental conditions. This predator can also be used year round due to the absence of diapause.

Lady beetles of the genus *Stethorus* are specialist predators of spider mites and are also useful for the control of *T. urticae* in greenhouses. A common species, *S. punctillum,* is a voracious predator and is commercially available. This species can consume over 1,000 spider mite eggs over a developmental span of two to three weeks in greenhouses. It is able to find small colonies of spider mites and has a very good dispersal ability. It is active and performs well within 33-90% RH and 20-30°C. In greenhouses, it can establish well on pepper and cucumber, but not on tomato. Because of their voracious appetite, these lady beetles are best released in 'hot spots' of spider mite infestation.

Some other generalist predators also feed on spider mites and may be of some use. Predatory Hemiptera used in biocontrol of whiteflies (e.g. *Macrolophus caliginosus*) and other insects are also useful biocontrol agents against spider mites to some degree. Lacewings are generalist predators and also used to control spider mites by periodic releases. A predatory ant, *Tapinoma melanocephalum* is known to attack *T. urticae* on *Salvia splendens* in central Florida greenhouses and has been shown to be a significant predator of *T. urticae.*

Thrips are generally considered to be harmful, but *Scolothrips sexmaculatus* attack spider mites and are used in biological control of spider mites. This species is adapted to hot and dry conditions. It is commercially available for spider mite control. Other thrips are also known as facultative predators of spider mites.

Entomophagous fungi may also be the biological control of twospotted spider mites. Two species, *Entomophthora thaxteriana* and *E. adjarica* can cause a heavy epizootic in populations of *T. urticae* and may be useful in greenhouses when humidity can be maintained near saturation for a period of time.

References & suggested further reading. Zilberminc *et al.* (1978); Egina and Cinovskij (1980); Sabelis (1981); Helle and Sabelis (1985b); Dong *et al.*

(1986); Jindra *et al.* (1991); Osborne *et al.* (1995); Fischer and Leger (1996); Sampson *et al.* (1996); Wilson *et al.* (1996); Gillespie *et al.* (1998); Rott and Ponsonby (2000a, b).

Chemical control

Chemical control of spider mites is becoming more and more difficult due to the rapid development of resistance in mites and the decrease of the number of registered acaricides for use. For example, clofentezine resistance in *T. urticae* was recorded in Australia in 1987 after mites in Queensland greenhouse roses had been exposed to 40 applications of clofentezine over a ten-month period; clofentezine resistance in this strain was extremely high (>2,500X) and conferred high level cross-resistance to the chemically unrelated compound hexythiazox. However, some chemicals appear to be less susceptible to resistance development in mites. *T. urticae* was subjected to four, six or 15 times of artificial selection of resistance to abamectin and no increased resistance to the pesticide was detected; abamectin is one of the most widely used chemicals in greenhouses. However, development of resistance of *T. urticae* to this pesticide has now been reported in some populations.

Chemicals and methods for control vary greatly depending on level of mite resistance, crop and environmental conditions, and the availability of registered material in different countries. No general recommendations will be given here. Interested readers should study the references suggested here and are advised to consult local pest control advisers in plant protection or pest control organizations.

References & suggested further reading. Green *et al.* (1984); Hoy and Conley (1987); Jensen and Mingochi (1988); Dong (1990); Richter and Schulze (1990); Papaioannu-Soulioti (1991); van de Veire and Degheele (1992); Aguiar *et al.* (1993); Herron *et al.* (1993); Szwejda (1993); Zhang and Sanderson (1995); Rossi and Conti (1997); Jacobson *et al.* (1999); Szwejda (1999); Makundi and Kashenge (2002).

Integrated control

More and more greenhouse pests are now under integrated pest management (IPM) programmes and *T. urtiace* is one of these. The key to the success of IPM programmes is careful monitoring of pest populations and the application of control measures only when necessary. That is, control measures are applied only when spider mite densities exceed the action threshold.

In IPM programmes, the use of pesticides is minimized and the use of biological control and other environmentally sound methods are encouraged. It is important that chemical control, when it is used, should be integrated with other control methods such as biological control.

Chemical control of *T. urticae* can be integrated with biological control by using selective chemicals that are less or not toxic to natural enemies, or by using chemicals in selected areas of the crop. In greenhouses in Belgium, for example, a stable equilibrium between *T. urticae* and the predator *P. persimilis* is achieved by selective use of chemicals and creating an asynchronous development in the predator population. The predators are introduced into one end of the greenhouse while acaricides (Torque 50% [fenbutatin oxide] and hexythiazox [Nissorun 10%]) are sprayed into the other end. Only 3,300 predators per 100 m^2 combined with three acaricide treatments applied to half the plants are required to control the pest for 30 weeks. Once the system is established, it is self-regulating and so the use of acaricides is needed only initially. Likewise, it may also be possible to apply chemical control to part of the plants while giving biological control a chance in other parts. For example, integrated control might be possible for spider mites on roses if different injury levels are assigned to the upper and lower canopies and sprays are confined to upper canopies. The upper portion of a rose canopy has an extremely low injury level because it bears the flowers and foliage which are cut for sale. Although low densities of spider mites may not affect the quantity of the product, they may damage its aesthetic appearance. However, the lower canopy can have a much higher injury level and aesthetic damage is of no concern on the lower canopy. Confining pesticide applications to only a portion of the canopy may create refugia for pesticide-susceptible individuals of *T. urticae* that may breed with resistant ones and thereby retard the development of pesticide resistance. This also allows *P. persimilis* to be used for biological control of spider mites on lower canopies.

Pesticide-resistant predatory mites can also be used in IPM and some strains of *P. persimilis* and *G. occidentalis* have been developed for controlling *T. urticae* in greenhouses.

Some plant cultivars are naturally less susceptible to spider mites and plant resistance may be used as a component in spider mite IPM. This will reduce the need for control measures. Sometimes, resistant plants have lower yield and other undesirable features, but plant breeding is helping to overcome some of these.

Cultural and physical methods may also be used. For example, spider mites thrive in hot, dry conditions. Increasing the humidity level by misting plants can reduce the growth of spider mites. However, care should be taken because increasing humidity may increase the possibility of fungal diseases.

References & suggested further reading. Field and Hoy (1986); Lindquist *et al.* (1987); Zhang and Sanderson (1990); Golovkina and Zvereva (1991);

van de Veire and Degheele (1992); Beck *et al.* (1993); Nihoul (1993); Smith *et al.* (1993); Gimenez-Ferreret *et al.* (1994); Park *et al.* (1995); Sanderson and Zhang (1995); Bennison *et al.* (1996); Sterk and Meesters (1997); Fejt and Jarosik (2000); Sacco *et al.* (2002).

4.4.2 *Tetranychus cinnabarinus* (Boisduval)

4.4.2.1 Common Name

This species is commonly known as the carmine spider mite. Informally, it is known by other names such as the red spider mite, the cotton spider mite and the carnation mite.

4.4.2.2 Distribution And Host Plants

This is a widespread species in subtropical areas of the world and common in greenhouses throughout temperate countries. It is less polyphagous than *T. urticae*, but it is difficult to tell from literature the true host range of this species because many authors do not separate *T. cinnabarinus* from *T. urticae*. In South Africa, for example, *T. cinnabarinus* attacks over 120 species of plants. This species attacks both vegetables (especially tomatoes, cucumbers, aubergines) and ornamentals (e.g. carnation, gerbera) in greenhouses.

References & suggested further reading. Jeppson *et al.* (1975); Meyer (1981, 1987); Bolland *et al.* (1998).

4.4.2.3 Appearance And Damage Symptoms

The eggs are laid singly on the under surface of leaves. They are spherical in shape and amber in colour, often with a distinct pale brownish spot or traces of red. Newly hatched six-legged larvae are yellow to orange in colour and become greenish after feeding. Eight-legged nymphs are yellowish green with dark spots, their bodies rounded in shape with short legs. Adult females are about 400-500 μm and males are smaller with a tapered hysterosoma. The females (summer form) are dark red in colour with two black spots on the dorsolateral idiosoma, with carmine colour extending to the eyes and the rest of the propodosoma is yellowish. Males are straw-coloured. Overwintering females are pale red or purple.

Symptoms caused by *T. cinnabarinus* vary slightly on different plants, but in general are very similar to symptoms caused by *T. urticae* (see 4.4.1.3). The carmine spider mite is often more phytotoxic than the twospotted spider mite, especially on tomato plants. The necrotic symptoms consist of premature chlorosis of infested leaflets which subsequently wither and die.

On tomatoes in the UK, some populations of greenish mites were recently found to be more similar to *T. cinnabarinus* than to *T. urticae* in both morphology and symptoms. Whether they are distinct or hybrids of the two species is being studied.

References & suggested further reading. Boudreaux (1956); Meyer (1981); Foster and Barker (1978); Zhang and Jacobson (2000).

4.4.2.4 Diagnostic Characters

Adult female (summer form): there is a diamond pattern of striae between setae e_1 and f_1 on dorsal hysterosoma (Fig. 4.1). The lobes on striae in the diamond area are often triangular in shape (cf: often rounded or semi-oblong and often wider than tall in *T. urticae*). The proximal tactile setae of tarsus I are proximal to the duplex setae as in Fig. 4.2 (cf: proximal tactile setae aligned with the proximal duplex setae in *T. ludeni*). Tibia I often has one to three solenidia in addition to ten tactile setae (cf: no solenidia in addition to ten tactile setae in *T. urticae*).

Adult male: empodium I is clawlike. The aedeagus is distinctive in having a small knob or head with its axis set at a small angle to the shaft axis; the anterior angulation is often slightly rounded, but the posterior angulation is acute (cf: the anterior and posterior angulations are both acute, small and equal in *T. urticae*); the dorsum of the knob is broadly angulated (cf: rounded in *T. urticae*).

References & suggested further reading. Boudreaux (1956); Brandenburg and Kennedy (1981); Meyer (1987); Kuang and Cheng (1990); Zhang and Jacobson (2000).

4.4.2.5 Life History And Biology

The life history and biology of this species are in general very similar to that of *T. urticae* (see 4.4.1.5). In fact, some papers on the biology of this species might have been reported under *T. urticae* because many authors consider *T. cinnabarinus* the red form of *T. urticae*.

Development is faster at higher temperatures and can be completed in less than a week at 35°C (Fig. 4.8). After a preoviposition period of one to two days, females on average lay 129 eggs at 24°C. On the same host species, the intrinsic rate of increase of *T. cinnabarinus* is lower than that of *T. urticae*. When they occur on the same plant, *T. cinnabarinus* is thus inferior in interspecific competition.

T. cinnabarinus shows very different performances on different plant species. Its oviposition rate on cucumber is several times higher than on pepper and tomato. On gerbera, the intrinsic rate of population increase is higher on cv. Ajax (0.18/day) than cv. Porto (0.14/day).

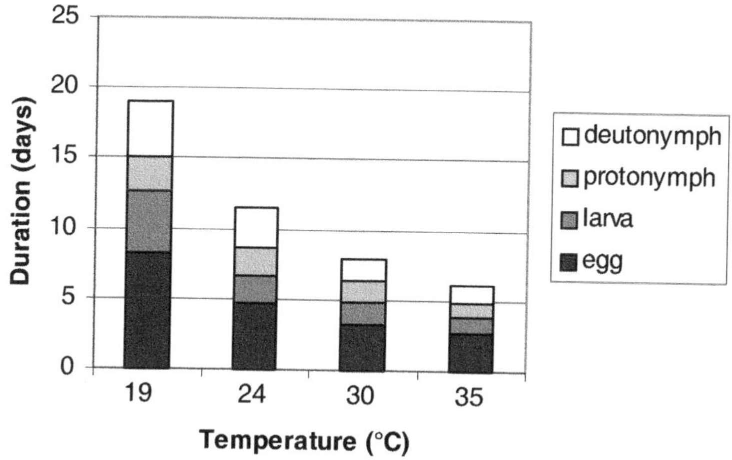

Fig. 4.8 *Tetranychus cinnabarinus*: developmental time from egg to adult in relation to temperature at RH of 80% (drawn from data in Hazan *et al.*, 1973)

Tetranychus cinnabarinus tends to be a pest of low-growing plants. Diapause is lost in many populations that remain on plants during winter. In diapausing populations, the induction and termination of diapause are controlled by day length. The critical photoperiod is about nine hours and 45 minutes. Both deutonymph and adult stages are photosensitive. In some strains, a period of chilling is not required and high temperature can reactivate feeding and oviposition.

References & suggested further reading. Davis (1961); Hazan *et al.* (1973); Hessein (1975); Vas Nunnes (1986); Northcraft and Watson (1987); Wu (1990); Gokkes *et al.* (1992); Witul (1992); Wu and Jing (1993); Kielkiewicz (1996); Kropezynska and Tomczyk (1996); Liu and Sun (1998); Witul and Kielkiewicz (1998); Bhagat and Singh (1999).

4.4.2.6 CONTROL AND MANAGEMENT

Control methods for *T. cinnabarinus* are similar to those for *T. urticae,* although on some plants (e.g. tomato), *T. cinnabarinus* is sometimes more difficult to control because they induce damage at lower population densities than *T. urticae.*

Phytoseiulus persimilis is the most effective predator of *T. cinnabarinus* and *T. urticae,* although some preference for *T. cinnabarinus* has been reported. *N. californicus* and *N. longispinosus* have also been used effetively against *T. cinnabarinus.* Detailed discussions on the use of phytoseiids on different crops are in Chapter 10.

Predatory insects are also effective natural enemies, including predatory midges (e.g. *Feltiella*), lady beetles (e.g. *Stethorus*), predatory Hemiptera (e.g. *Orius*), predatory thrips (e.g. *Scolothrips*). Spiders are also considered significant predators. Entomophagous fungi (e.g. *Hirsutella thompsonii*) can also be effective if humidity can be maintained very high for a period of time and temperature is lower than 37°C.

Chemical and other control methods are also similar to those used against *T. urticae*. In the interest of space, no further discussions will be given here. Interested readers should read references suggested here and are advised to consult pest control advisers in plant protection or pest control organizations.

References & suggested further reading. Foster and Barker (1978); Gerson *et al.* (1979); Berlinger *et al.* (1988); Dong (1990); Szwejda (1993); Mansour *et al.* (1995); Bennison *et al.* (1996); Gu *et al.* (1996); Kazak *et al.* (1997); Ho and Chen (1998); Valunj *et al.* (1999); Colkesen and Sekeroglu (2000); Edelstein *et al.* (2000); Kazak *et al.* (2000); Walzer and Schausberger (2000); Schausberger and Walzer (2001); Karaca *et al.* (2002).

4.4.3 Other spider mites

4.4.3.1 *Tetranychus ludeni* Zacher

Commonly known as the dark-red spider mite, red-legged spider mite or bean mite, this species is widespread in the tropics and has been recorded from over 300 species of plants worldwide. It is a serious pest of bean, eggplant, hibiscus, pumpkin and other cucurbitaceous plants in warm areas. It is also quite common on greenhouse plants in temperate areas.

The eggs are often laid on the under surface of leaves in most plants. They are spherical in shape and pallid yellow or darker in colour. Six-legged larvae and eight-legged nymphs are darker in successive stages. The adults are dark red in colour but without two black spots on dorsolateral idiosoma as seen in twospotted spider mites. Legs are often dark red as the name suggests. This species is very similar to *T. cinnabarinus* in general appearance because of the carmine colour. Morphologically, the males of *T. ludeni* can be easily distinguished from those of *T. cinnabarinus* by the absence of the posterior angulation of the knob in the male aedeagus (in *T. cinnabarinus*, both anterior and posterior angulations are present; Fig. 4.5). Females of *T. ludeni* can be distinguished from those of *T. cinnabarinus* by the alignment of proximal tactile setae in tibia I with the proximal pair of duplex setae (cf: in *T. cinnabarinus*, the proximal tactile setae in tibia I are distal to the proximal pair of duplex setae; Fig. 4.5).

Symptoms caused by *T. ludeni* are different on different plants, but in general are similar to symptoms caused by *T. urticae* and *T. cinnabarinus* (see 4.4.1.3). Attacked eggplants show a yellowish hue; leaves wilt and drop as mite feeding continues. Damage is more severe in dry conditions. Moderate populations may greatly reduce crop yield and heavy infestation can kill host plants.

Adults produce profuse webbing, which provides a suitable microhabitat within which feeding and reproduction are concentrated. They also deposit black and white fecal pellets in the colony. These pellets are believed to act as regulators of ambient relative humidity in the microhabitat.

This is a tropical and subtropical species. Development occurs above 14.7°C. On beans (*Phaseolus vulgaris*), the mean duration of each immature stage is: egg 4.7 days, larva 1.8 days, protonymph 1.3 days and deutonymph 1.9 days (at 26±4°C). The net reproduction rate is 77.4, mean generation time 19.6 days, intrinsic rate of natural increase 0.253 individuals/female/day and finite rate of natural increase 1.287 individuals/female/day. As expected, development is faster at higher temperatures; at 35°C, development from the egg to adult can be completed within one week (Fig. 4.9).

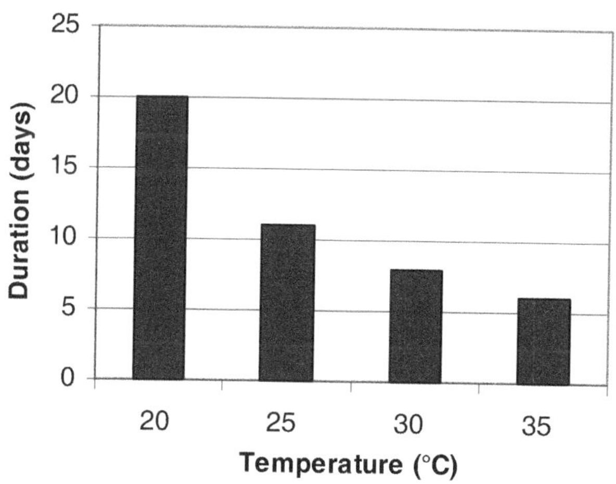

Fig. 4.9 *Tetranychus ludeni*: developmental time from egg to adult in relation to temperature (drawn from data in Silva *et al.*, 1999).

There have been few reports on the biological control of this species. In Australia, *T. ludeni* often occurs together with *T. urticae* in strawberries but both can be controlled by *P. persimilis*. Several other species of phytoseiids are known to prey on this species but their effects are not known: *Typhlodromips tetranychivorus* Gupta, *Paraphytoseius multidentatus* Swirski and Shechter, *N. fallacis* (Garman) and *N. longispinosus* (Evans).

The predatory mite *Cunaxa setirostris* (Hermann) (of the family Cunaxidae) feeds on active stages of *T. ludeni*. One female predator consumes an average of 330 mobile prey during its life span. The webbing produced by *T. ludeni* has no negative effect on the number of prey consumed by *C. setirastris*. This species has good potential as a biological control agent against *T. ludeni*.

Two species of lady beetles (*Stethorus pauperculus* and *Scymnus coccivora* Ramakrishna) and a rove beetle (*Oligota oviformis*) are also predators of *T. ludeni*, but their effectiveness is not known.

Sulphur, dicofol, tetradifon and seed kernel extract of neem (*Azadirachta indica*) are effective against *T. ludeni*. Readers interested in chemical control should read references suggested here and are advised to consult local pest control advisers in plant protection or pest control organizations. Whenever possible, use common sense pest control methods, and chemicals as the last option (Table 1.5).

References & suggested further reading. Jeppson *et al*. (1975); Biasi and Santos (1988); Goodwin (1990); Ansari and Pawar (1992); Kumar and Sharma (1993); Morros and Aponte (1994); Sumangala and Haq (1994); Bolland *et al*. (1998); Silva *et al*. (1999); Arbabi and Singh (2000); Waite (2001).

4.4.3.2 *Tetranychus kanzawai* Kishida

This species is commonly known in Asia as the kanzawa spider mite. It is also known as the tea red spider mite because its main host is tea.

This species is an important pest throughout East and Southeast Asia (mainland China, Hong Kong, Taiwan, Korea, Japan, Okinawa Island, India, Thailand, Malaysia, Indonesia and Philippines) and is now also known to occur in Australia, Papua New Guinea, South Africa, Congo, Colombia, Mexico, the USA and Greece. It attacks over a hundred species of plants, including many crops and ornamental plants. It is normally an outdoor species, but can attack greenhouse plants as well. It is a pest of grapes in greenhouses in Japan and of *Angelica utilis* in Korea.

The eggs are often laid on the under surface of leaves. They are spherical in shape and are clear when first laid. Six-legged larvae and eight-legged nymphs are yellowish green. The adults are red or yellowish red, depending on host plants. Legs are not red but yellowish. Morphologically, this species is very similar to *T. cinnabarinus*. The males of *T. kanza-*

wai can be distinguished from those of *T. cinnabarinus* by its larger knob in the male aedeagus (Fig. 4.5).

Symptoms caused by *T. kanzawai* are different on different plants, but in general are similar to symptoms caused by *T. urticae* and T. *cinnabarinus*.

Development occurs above 10.3°C. Development time from egg to adult is 19, 16 and 12 days at 20, 22 and 25°C, respectively. Survival to the adult stage is about 80%. Adult sex ratio is female biased and averages 67%. Adult life span is 20-33 days in females and 19-35 days in males at 15-30°C. Fecundity ranges from 28 eggs/female at 15°C to 76 at 30°C. At 30°C, the intrinsic rate of increase is 0.276, the net reproductive rate 31.1, the finite rate of increase 1.318 and the mean generation time 12.4 days. Reproductive diapause is facultative and a higher proportion of females enters diapause at lower temperatures; 100% of females enter diapause at 16°C.

The kanzawa spider mite can be controlled by phytoseiid predators. In Japan, *P. persimilis* provides effective control of *T. kanzawai* populations on grapes grown in greenhouses when predators are introduced in June, and pesticides with a long-term toxicity are not sprayed before the introduction of the predators. In Taiwan, both *N. fallacis* and *P. persimilis* give effective control of *T. kanzawai* after introduction onto potted tea plants at a predator:prey ratio of 1:20 in a greenhouse. In Korea, *T. kanzawai* is controlled by *P. persimilis* on *Angelica utilis* in 'walk-in' plastic tunnels. *N. longispinosus* is also commonly associated with *T. kanzawai*, but its effectiveness is affected by temperature and humidity. At relative humidities of less than 70%, the hatchability of *N. longispinosus* eggs is reduced. Between 20-30°C and 70-100% RH, the effectiveness of this predator increases with temperature and RH.

Several species of predatory insects are also voracious predators of *T. kanzawai*. The rove beetle *Oligota flavicornis* consumes 160 spider mite eggs per day and the predatory thrips *Scolothrips indicus* consumes 80 spider mite eggs. These predators can be used in hot spots of spider mites. Some anthocorid species of *Orius* are efficient at suppressing *T. kanzawai* populations in unsprayed aubergines. The lacewing *Mallada basalis*, released as eggs or larvae on every strawberry plant at three-week intervals kills 60 to 90% of a *T. kanzawai* population. On greenhouse grapes in Japan, several other predators are also found feeding on this mite: predatory midges *Feltiella* sp., predatory thrips *Scolothrips takahashii* and the spiders *Oligota kashmirica benefica* and *Achaearanea tepidariorum*.

Chemical and other control methods are similar to those for *T. urticae* and are not further discussed here. Interested readers should study references suggested here and are advised to consult local pest control advisers in plant protection or pest control organizations.

References & suggested further reading. Jeppson *et al.* (1975); Tsai *et al.* (1989); Ho (1990); Yamada and Tsutsumi (1990); Nakagawa (1991); Yang *et al.* (1991); Ashihara *et al.* (1992); Fujibayashi and Sekita (1993); Takafuji and Inoue (1993); Ashihara (1995); Chang and Huang (1995); Kim and Lee (1996); Bolland *et al.* (1998); Kim *et al.* (1999); Ho and Chen (2001, 2002).

4.4.3.3 *Tetranychus turkestani* (Ugarov & Nikolshi)

This species is most commonly known as the strawberry spider mite or simply the strawberry mite, and in the USA the McGregor strawberry mite becasue he named *Tetranychus atlanticus*, a synonym of this species.

This is a widespread species and has been known from Europe (Bulgaria, CIS, Canary Islands, France, Greece, Hungary, Poland, Portugal, Spain, Switzerland, The Netherlands, Turkey, Yugoslavia), Asia (China, Iran, Iraq, Japan, Kuwait, Pakistan), North and Central America (Costa Rica, Mexico, the USA), Africa (Algeria, Morocco, Israel, South Africa) and New Zealand. This species is a very polyphagous species, known from over 180 species of plants. It is a serious pest of low-growing crops such as beans, cucumber, eggplant, melons and strawberry. It is occasionally found on greenhouse plants (e.g. cucumbers) in Eastern Europe.

The eggs are often laid on the under surface of leaves. They are spherical in shape and are clear when first laid, becoming opaque and finally ivory before hatching. Six-legged larvae are pale at first and then become greenish with two black spots on the dorsal idiosoma after feeding. The eight-legged nymphs are pale straw-coloured with two large black spots on dorsal idiosoma. The adults are green, amber, brownish or almost black, depending on host plants; there are two large black spots on each side from the eyes to mid-dorsum. Morphologically, this species is very similar to *T. urticae*. The males of *T. turkestani* can be distinguished from those of *T. urticae* by its larger knob or head of the aedeagus and its rounded anterior angulation (Fig. 4.5).

The upper surface of infested leaves shows dead areas where *T. turkestani* feed in colonies on the lower surface. High mite densities produce sufficient webbing to cause the leaves and stems to become matted together. Heavy mite damage causes leaf drop and even the death of plants. Immature stages produce as much injury as adult females, but adult males produce no visible injury.

Development from egg to adult takes 11 days at 25°C. Each female lays up to nine eggs per day and a total of about 85 eggs during the entire life span. Sex ratio is only slightly female biased, with 55.7% daughters. Short day-length induces reproductive diapause and the mite colour gradually changes to bright orange.

Phytoseiid mites have not been studied for the control of this species in greenhouses, but many effective predators of *T. urticae* (e.g. *P. persimilis*) should be applicable to this species.

The generalist predator *Anystis baccarum* (L.) (Anystidae) feeds on all stages of *T. turkestani*. Several predatory insects are also known to feed on this species: coccinellids *Stethorus gilvifrons* and *Exochomus pubescens*, the chrysopid *Chrysoperla mutata* and the thrips *Scolothrips sexmaculatus*.

The species is known to be susceptible to sulphur. Chemicals effective against other *Tetranychus* mites may be applicable to this species, but use common sense pest control methods first and chemicals as the last option (Table 1.5).

References & suggested further reading. Jeppson *et al.* (1975); Carey and Bradley (1982); Popov (1988); Al-Mallah and Abdalla (1990); Bolland *et al.* (1998); Ahmed and Ahmed (1989); Khanjani *et al.* (1999).

4.4.3.4 *Eotetranychus lewisi* (McGregor)

This species is known as the lewis spider mite or simply the lewis mite. It is also known as the poinsettia spider mite because it is the most significant mite pest of greenhouse poinsettia.

This species is widely distributed from North to South America (Bolivia, Chile, Colombia, Costa Rica, El Salvador, Guatemala, Honduras, Mexico, Nicaragua, Panama, Peru and the USA) and is also known from the Madeira Islands, South Africa and Libya. It is known to feed on over 60 plant species including fruit crops (citrus in California and papaya in Mexico) and ornamental flowers (poinsettia in the USA and Canada).

The damage symptoms these mites inflict on crops is similar to those of *T. urticae*. They feed on the under surface of leaves, piercing the epidermis and removing cell contents. This produces a speckled or peppered effect on the foliage. Eventually the entire leaf becomes chlorotic and may fall off. With continued feeding or heavy infestations, loose webbing is produced, under which the mite colony lives.

The eggs are spherical in shape and are laid in depressions on leaf surfaces. They have a short stalk with no radiating guy fibrils. The adult female is oval in shape and about 360 μm long. It is pale-greenish at first but becomes amber with ageing. There are a varying number of black spots along lateral margins, but usually one pair above leg III and another pair near the posterior end. The peritremes are hooked distally. The striation pattern is transverse on the genital flap and on the area anterior to it.

Tarsus I has five tactile setae proximal to the duplex setae. Tibia I and II have eight and nine tactile setae, respectively. Males are smaller than females, wedged-shaped, and mustard-yellow in colour. The aedeagus is gradually tapered distally and forms a broad sigmoid bend (cf: curved dorsad near the middle of the shaft, but the distal portion is directed caudoventrally with a deflexed tip in *E. sexmaxculatus*; Fig. 4.4).

The developmental time from egg to adult is about two weeks for the females. Males mature a couple of days earlier than females because they have only one nymphal stage. Females lay two to three eggs per day for about 30 days. This species reproduces continuously without diapause.

Biological control of this mite has not been studied but many commercially available predatory mites (e.g. *N. californicus*) or insect predators (coccinellid beetles) that are used against other mites of Tetranychinae may be applicable to this species.

This species is susceptible to most registered miticides, but application should be carried out before leaves begin dropping off the plants. Whenever possible, use common sense pest control methods, and chemicals only as the last option (Table 1.5).

References & suggested further reading. McGregor (1943); Doucette (1962); Jeppson *et al.* (1975); Bolland *et al.* (1998).

4.4.3.5 *Eotetranychus sexmaculatus* (Riley)

This species is known as the sixspotted spider mite or sometimes simply the sixspotted mite.

This species is distributed in Asia (China, India, Iraq, Japan, Korea, Okinawa Island, Taiwan), Oceania (Australia and New Zealand), North America (the USA) and South America (Chile). It has been reported from over 30 plant species, including crops (citrus, avocado and grape) and ornamentals (azalea). It is a periodic pest of grape vines in New Zealand and is occasionally found in greenhouses.

This species forms colonies covered in webbing on the under surface of leaves. Mite feeding produces yellow depressions on the under surface and raised yellow or yellowish white spots on the upper surface. As mite feeding and infestation continue, yellow spots merge and the leaf becomes distorted and drops prematurely.

The eggs are spherical in shape and transparent or pale greenish yellow in colour. They have a short stalk with no radiating guy fibrils. The adult females are about 300 μm long and oval in shape. They are lemon yellow in colour, often with six black spots on the dorsal idiosoma. The peritremes are hooked distally. The striation pattern is longitudinal on the anterior central portion of the genital flap and on the area anterior to it (cf:

transverse in *E. lewisi*). Tarsus I has five tactile setae proximal to the duplex setae. Tibia I and II have eight and nine tactile setae, respectively. Males are smaller than females and wedge-shaped. The aedeagus is slightly curved dorsad near the middle of the shaft, but the distal portion is directed caudoventrally with a deflexed tip (Fig. 4.4).

Development from egg to adult takes about 11 to 20 days in summer. After a preovipostion period of two to three days, females start to lay two to three eggs per day, with a total of 25 to 40 eggs per female. Eggs require five days to three weeks to hatch, depending on temperature. Unlike *T. urticae*, this species is adversely affacted by dry conditions; it thrives in warm, humid conditions favoured by many phytoseiids.

Because of its "climate-match" with phytoseiids, it is not surprising that this species is generally kept under control by phytoseiid mites. Phytoseiid species known to feed on this species include: *Galendromus helveolus, G. occidentalis, G. porresi, G. annecten, N. californicus, N. collegae, Euseius hibisci*, and *E. mesembrinus*. This species periodically becomes a pest when predators are disrupted (e.g. by pesticides).

When it becomes a problem in greenhouses, reducing the humidity level can reduce mite growth. Use of sulphur or oil sprays can also control the mite. This species is susceptible to most registered miticides, but application should be done before leaves begin dropping off the plants. Whenever possible, use common sense pest control methods, and chemicals only as the last option (Table 1.5).

References & suggested further reading. Jeppson *et al*. (1975); Abou-Setta and Childers (1989); Caceres and Childers (1991); Mizell and Schiffhauer (1991); Bolland *et al*. (1998).

4.4.3.6 *Panonychus citri* (McGregor)

This species is commonly known as the citrus red mite. It is of worldwide distribution and known from over 80 species of plants. It is a major pest of citrus and occasionally attacks grapes, ornamental flowers and evergreen shrubs grown in greenhouses and nurseries.

The eggs are pale to red, nearly spherical, somewhat flattened on the bottom and stalked on the top with guy fibrils radiating from the tip of the stalk to the leaf surface. Three-legged larvae and eight-legged nymphs and adults are dark red to purplish in colour. The adults have dorsal setae arising from red tubercles (cf: white tubercles in *P. ulmi*, another common species of the genus). The striae on the genital plate are transverse but those anterior to the plate are longitudinal. On dorsal hysterosoma, the clunal setae are similar in length to outer sacral setae, both about one-third as long as the inner sacral (cf: the clunal setae about two-thirds of the outer sacral setae, which are more than half as long as the inner sacral in *P. ulmi*).

This species feeds on the upper surface of leaves and produces a stippled appearance initially, which develops into pale patches later. With continued feeding and damage, the leaves become grey, silver or yellow.

Development occurs above 10°C and all stages die at 40°C. Developmental time is shorter at higher temperatures and is about ten days near 25°C (Fig. 4.9), which is the optimal temperature. At this temperature, adult females live for about nine days and lay an average of 25 eggs. Sex ratio is biased, with about 70% females. A relative humidity of 65% is optimal for development and reproduction. This species prefers moderate climatic conditions. Low humidity and very high temperature are detrimental to population development of this mite.

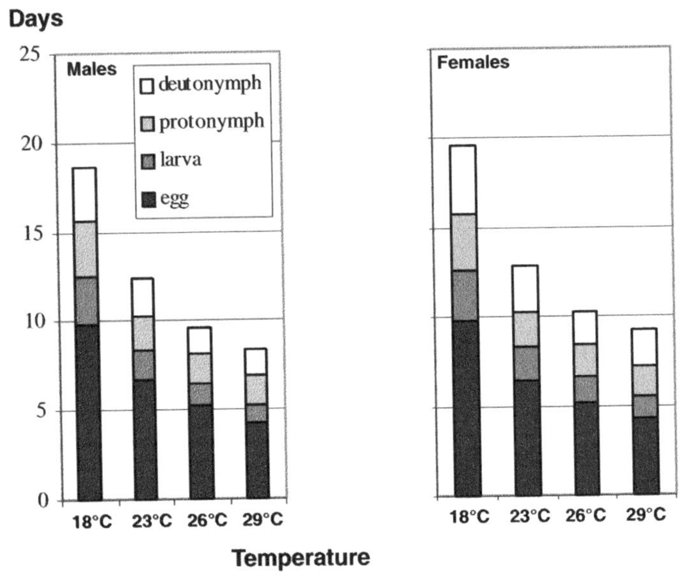

Fig. 4.9 *Panonychus citri*: developmental time from egg to adult in relation to temperature for males and females (drawn from data in Tian and Pang, 1997).

This species seems to be a pest induced by pesticides. In unsprayed crops, they are often naturally controlled by a complex of predators including predatory mites and predatory insects.

In the Willamette Valley, Oregon, USA, *P. citri* attacks and causes damage to an ornamental plant, *Skimmia japonica*, in nurseries as well as outdoor landscape, but *Neoseiulus fallacis*, when released, can provide effective control of this mite. In Cuba, *Phytoseiulus macropilis* released at 1:20 predator:prey ratios in orange cv. Valencia plants grown in nurseries can maintain prey at low levels. *Neoseiulus cucumeris*, which is easy and not expensive to produce, also feed on citrus red mites and can provide provide effective control when released against this mite. Several other species of phytoseiids have been found to be natural enemies of citrus red mites. In addition to phytoseiids, a stigmaeid mite (*Agistemus exsertus*) has been demonstrated effective in the control of *P. citri* in a greenhouse release study.

Predatory insects are also effective predators of citrus red mites. They include: *Stethorus* beetles (e.g. *S. punctillum*), predatory thrips (e.g. *Scolothrips takahashii*), and lacewings (e.g. *Mallada basalis*, *Chrysoperla sinica* and *Propylea japonica*).

This species has developed resistance to many pesticides and some chemicals (e.g. pyrethroids) can stimulate an outbreak of the pest. Please consult pest control advisers in plant protection or pest control organizations.

Some organic or soft chemicals may be used. The seed oil of chinaberry (*Melia azedarach*), for example, gives 94-96% control of *P. citri* on citrus when sprayed with a 0.5% seed oil emulsion. It is not phytotoxic and is also harmless to the natural enemies.

References & suggested further reading. Jeppson *et al.* (1975); Luo *et al.* (1988); Wei *et al.* (1989); Furuhashi (1990); Vierbergen (1990); Beitia and Garrido (1991a,b); Zhou *et al.* (1991); Huang (1994); Yue and Tsai (1995); Wu and Wu (1995); Tian and Pang (1997); Bolland *et al.* (1998); Pratt and Croft (1998); Zhang *et al.* (2001); Ramos and Santos (2002).

4.4.3.7 *Oligonychus perditus* Pritchard and Baker

This species is distributed mainly in China (including Hong Kong and Taiwan), Japan and Korea, but also in The Netherlands and the USA. It has been recorded from some 15 plant species (esp. cupressaceous conifers). There is only one report of this species from bonsai *Juniperus chinensis* in greenhouses in The Netherlands.

This species is a member of the *ununguis*-group of the genus *Oligonychus*, which is characterized by having seven tactile setae on tibia I (plus one solenidion in female and four solenidia in male). This species is characterized by having two ventral tactile setae beyond the duplex setae on tarsus I.

Little is known about its biology and control, although it is considered a quarantine species.

References & suggested further reading. Jeppson *et al.* (1975); Bolland *et al.* (1998); Vierbergen (1990).

4.4.3.8 *Bryobia* species

Species of *Bryobia* are mainly pests of outdoor plants. A few species are known to sometimes invade sheds, greenhouses or other buildings and cause some damage to plants or nuisance to people.

The eggs of *Bryobia* are spherical and smooth. They are slightly sticky and are often covered with fine dust particles. Eggs are laid on the under surface of leaves and are red in colour. Larvae are often bright red at first but after feeding become green and almost spherical. The protonymphs and deutonymphs are brownish at first but after feeding become greenish.

Bryobia mites are often found on low-growing plants. Unlike *Tetranychus* mites, *Bryobia* mites enter diapause as eggs in most species, prefer upper surfaces of leaves, and do not produce webbing.

Phytoseiid mites are the major natural enemies of these mites but there have been few studies on their use. *Bryobia* mites are conventionally controlled by acaricides when they become a problem. Many pesticides are known to be effective against these mites. Development of pesticide resistance has rarely been reported. Among the many chemicals used is 0.2% dicofol, which seems still effective. Two fungicides, dichlofluanid and triforine, can also provide effective control of *Bryobia* without harming the natural enemies. Oil sprays are also known to be effective in some cases.

Four species of *Bryobia* have been known in greenhouses and they can be identified using the key in Fig. 4.4.

Bryobia kissophila van Eyndhoven. This species is known as the ivy mite. It is distributed mainly in Europe and also found in Chile, Costa Rica, Tasmania of Australia, and New Zealand. It is rather specific to ivy (*Hedera*), with only one other reported host (*Solanum tuberosum*). In Europe, this species occasionally causes damage to ivy grown in greenhouses.

Development occurs above 0°C and there may be six to eight successive generation in a year, with no true diapause. There are no males in this species.

Bryobia lagodechiana Reck. This species is distributed in CIS, Hungary, Poland, Switzerland, The Netherlands, Canada, New Zealand and Japan. It feeds on over 20 species of plants and attacks cucumber, beans and eggplant in greenhouses in the Ukraine. However, it does not cause serious damage to these plants because of its slow development and low reproductive rates.

Phytoseiid predators show promise as biological agents. Females of *Neoseiulus herbarius* (= *Neoseiulus graminis*) consume up to 12 eggs and/or early-instar nymphs of this species per day at 26°C, but they avoid large nymphs and adult prey. Females of *Phytoseiulus persimilis* also can consume up to seven eggs or larvae of this species and lay one to two eggs per day.

Bryobia praetiosa (Koch). This species is known as the brown clover mite or simply the clover mite. This is a cosmopolitan species known from over 250 species of plants. It may be of some economic importance on clover, lucerne, lawns, cereals and flowers.

The clover mite overwinters as eggs and other stages but oversummers in eggs only. The overwintering mites often move to protected places to molt or lay eggs. As a result, vast numbers of *B. praetiosa* may occasionally invade glasshouses/dwellings and become a nuisance, which may be controlled by a grass-free strip of one to three m outside of the greenhouse and treatment of the remaining grass with dicofol or chlorobenzilate.

An undetermined species of the *B. praetiosa* group is known to cause damage to Brussels sprout seedlings and another species to cauliflower seedlings in greenhouses in the UK. On Brussels sprout seedlings, these mites feed on the cotyledons and produce discrete lines of punctures on the first true leaves. Later, small necrotic feeding patches develop into distinct holes at the apical end of these leaves.

Bryobia calida Karg. This species was recently described from females collected from greenhouse cucumbers in Germany. Nothing more is known about this species.

References & suggested further reading. Evans *et al.* (1961); Hussey *et al.* (1969); Gordon (1975); Akimov and Kolodochka (1985); Bassett (1985); Hussey (1985); Karg (1985); Kolodochka (1985a,b); Wilkin and Warner (1985); Alford (1994).

4.5 References cited

Abou-Setta, M.M. and Childers, C.C. (1989) Biology of *Euseius mesembrinus* (Acari: Phytoseiidae): life tables and feeding behavior on tetranychid mites on citrus. *Environmental Entomology* 18, 665-669.

Aguiar, E.L., Carvalho, G.A., Menezes, E.B. and Machado, C.A. (1993) Efficacy of the acaricide/insecticide diafentiuron in the control of the two-spotted spider mite *Tetranychus urticae* (Koch) on roses. *Anais da Sociedade Entomologica do Brasil* 22, 577-582.

Ahmed, Z.I. and Ahmed, R.F. (1989) Biological studies of predator *Stethorus gilvifrons* Mulsant. (Coleoptera, Coccinellidae) on the strawberry mite *Tetranychus turkestani* Ugarov and Nikolski (Acariformes, Tetranychidae). *Journal of Biological Science Research* 20, 33-22.

Akimov, I.A. and Kolodochka, L.A. (1985) The mite *Bryobia lagodechiana* Reck (Trombidiformes, Bryobiidae) on cucumber plants. *Vestnik Zoologii* 1985(1), 45.

Alford, D.V. (1994) *A Colour Atlas of Pests of Ornamental Trees, Shrubs and Flowers.* Wolfe Publishing Ltd., London. 448 pp.

Al-Mallah, N.M. and Abdalla, S.I. (1990) On the biology of the strawberry mite *Tetranychus turkestani* Uga. & Nik. (Tetranychidae, Acariformes) and the susceptibility of three gladiolus cultivars to infestation under greenhouse conditions. *Arab Journal of Plant Protection* 8, 21-24.

Ansari, M.A. and Pawar, A.D. (1992) Biology of spider mite, *Tetranychus ludeni* Zacher (Acari: Tetranychidae) recorded on waterhyacinth. *Plant Protection Bulletin Faridabad* 44, 28-31.

Arbabi, M. and Singh, J. (2000) Studies on biological aspects of predaceous mite *Cunaxa setirostris* on *Tetranychus ludeni* at laboratory condition in Varanasi, India. *Journal of Agriculture and Rural Development* 2(1-2), 13-23.

Ashihara, W. (1995) Studies on the occurrence of the kanzawa spider mite, *Tetranychus kanzawai* Kishida (Acarina: Tetranychidae) and its biological control by *Phytoseiulus persimilis* Athias-Henriot (Acarina: Phytoseiidae) on grapevine in greenhouses. *Bulletin of the Fruit Tree Research Station* Extra 1995, No. 6, 75-151.

Ashihara, W., Inoue, K., Osakabe, M. and Hamamura, T. (1992) Effectiveness of *Phytoseiulus persimilis* Athias-Henriot (Acarina: Phytoseiidae) as a control agent for the Kanzawa spider mite, *Tetranychus kanzawai* Kishida (Acarina: Tetranychidae) and occurrence of native natural enemies of the spider mite on grapevine in glasshouse. *Bulletin of the Fruit Tree Research Station* 22, 109-130.

Baker, E.W. and Tuttle, D.M. (1994) *A Guide to the Spider Mites (Tetranychidae) of the United States.* Indira Publishing House, West Bloomfield, Michigan. 347 pp.

Bassett, P. (1985) *Bryobia* - the clover mite. In: Hussey, N.W. and Scope, N.E.A. (1985) *Biological Pest Control. The Glasshouse Experience*. Blandford Press, Dorset. pp. 91-92.

Beck, N.G., Workman, P. and Martin, N. (1993) IPM for Cymbidium orchids in New Zealand. *Bulletin OILB SROP* 16(8), 12-15.

Beitia, F. and Garrido, A. (1991a) Influence of relative humidity on development and egg-laying in *Panonychus citri* under controlled conditions. *Bulletin OEPP* 21, 719-722.

Beitia, F. and Garrido, A. (1991b) Influence of temperature on development and oviposition in *Panonychus citri* (McGregor) (Acarina: Tetranychidae), in controlled conditions. *Investigacion Agraria, Produccion y Proteccion Vegetales* 6, 255-262.

Bennison, J.A., Sampson, C., Vautier, A., Challinor, P.F. and Lenteren, J.C. van (1996) Development of IPM on protected aubergine. *Bulletin OILB SROP* 19, 7-10.

Berlinger, M.J., Dahan, R. and Mordechi, S. (1988) Integrated pest management of organically grown greenhouse tomatoes in Israel. *Applied Agricultural Research* 3(5), 233-238.

Bhagat, K.C. and Singh, W. (1999) Some observations on the biology and behaviour of carmine spider mite *Tetranychus cinnabarinus* (Boisduval) (Acarina: Terranychitae) - a pest of brinjal vegetable. *Journal of Advanced Zoology* 20, 28-31.

Biasi, L.A. and Santos, H.R. de (1988) Consumption of eggs of *Tetranychus ludeni* Zacher, 1913 (Acari: Tetranychidae) by *Amblyseius fallacis* German, 1948 (Acari: Phytoseiidae). *Revista do Setor de Ciencias Agrarias* 10(1-2), 99-102.

Bolland, H.R., Gutierrez, J. and Flechtmann, C.H.W. (1998) *World Catalogue of the Spider Mite Family* (Acari: Tetranychidae). Leiden, Brill. 392 pp.

Bosse, T.C. and Veerman, A. (1996) Involvement of vitamin A in the photoperiodic induction of diapause in the spider mite Tetranychus urticae is demonstrated by rearing an albino mutant on a semi-synthetic diet with and without beta-carotene or vitamin A. *Physiological Entomology* 21, 188-192.

Boudreaux, H.B. (1956) Revision of the two-spotted spider mite (Acarina, Tetranychidae) complex, *Tetranychus telerius* (Linnaeus). *Annals of the Entomological Society of America* 49, 43-48

Brandenburg, R.L. and Kennedy, G.G. (1981) Differences in dorsal integumentary lobe densities between *Tetranychus urticae* Koch and *Tetranychus cinnabarinus* (Boisduval) (Acarina: Tetranychidae) from northeastern North Carolina. *International Journal of Acarology* 7, 231-234.

Caceres, S. and Childers, C.C. (1991) Biology and life tables of *Galendromus helveolus* (Acari: Phytoseiidae) on Florida citrus. *Environmental Entomology* 20, 224-229.

Carey, J.R. and Bradley, J.W. (1982) Developmental rates, vital schedules, sex ratios and life tables for *Tetrancyhus urticae, T. turkestani* and *T. pacificus* (Acarina: Tetranychidae) on cotton. *Acarologia* 23, 333-345.

Chang, C.P. and Huang, S.C. (1995) Evaluation of the effectiveness of releasing green lacewing, *Mallada basalis* (Walker) for the control of tetranychid mites on strawberry. *Plant Protection Bulletin Taipei* 37, 41-58.

Colkesen, T. and Sekeroglu, E. (2000) The effect of *Amblyseius longispinosus* (Evans) (Acarina: Phytoseiidae) on *Tetranychus cinnabarinus* Boisd. (Acarina: Tetranychidae) on different cucumber cultivars. *IOBC/WPRS Bulletin* 23(1), 187-194.

Davis, D.W. (1961) Biology of *Tetranychus multisetis* the polychaetous form of *T. cinnabarinus*. *Annals of the Entomological Society of America* 54, 30-34.

Dong, H.F. (1990) Control effect of the enhanced Liuyangmycin on leaf spider mites and its influence on the beneficial phytoseiid predators. *Chinese Journal of Biological Control* 6, 97-101.

Dong, H.F., Guo, Y.J. and Niu, L.P. (1986) Biological control of the two spotted spider mites with *Phytoseiulus persimilis* on four ornamental plants in greenhouses. *Chinese Journal of Biological Control* 2, 59-62

Doucette, C.F. (1962) The lewis mite, *Eotetranychus lewisi* on greenhouse poinsettia. *Journal of Econimic Entomology* 55, 139-140.

Edelstein, M., Tadmor, Y. and Mansour, F. (2000) Resistance of *Cucurbita* and *Lagenaria* to the carmine spider mite, *Tetranychus cinnabarinus* (Acari: Tetranychidae). *Acta Horticulturae* 510, 283-288.

Egina, K.Ja. and Cinovskij, Ja.P. (1980) Results of evaluation of efficacy of a fungus *Entomophthora thaxteriana* (Petch) Hall *et* Bell on aphids and spider mites after storage. In: Cinocskij, Ja.P. (ed.) *Biologiceskij Metod Borby v Vrednymi Nasekomymi*. Zinatne, Riga, pp. 7-11.

Evans, G.O., Sheals, J.G. and Macfarlane, D. (1961) *The Terrestrial Acari of the British Isles. An Introduction to Their Morphology, Biology and Classification.* British Museum, London, 219 pp.

Fejt, R. and Jarosik, V. (2000) Assessment of interactions between the predatory bug *Orius insidiosus* and the predatory mite *Phytoseiulus persimilis* in biological control on greenhouse cucumber. *Plant Protection Science* 36, 85-90.

Field, R.P. and Hoy, M.A. (1986) Evaluation of genetically improved strains of *Metaseiulus occidentalis* (Nesbitt) (Acarina: Phytoseiidae) for integrated control of spider mites on roses in greenhouses. *Hilgardia* 54(2), 1-32.

Fischer, S. and Leger, A. (1996) *Macrolophus caliginosus* W. (Heteroptera, Miridae), a beneficial to be exploited under glass. *Revue Suisse de Viticulture, d'Arboriculture et d'Horticulture* 28, 111-112.

Foster, G.N. and Barker, J. (1978) A new biotype of red spider mite (*Tetranychus urticae* Koch) causing atypical damage to tomatoes. *Plant Pathology* 27, 47-48.

Fujibayashi, Y. and Sekita, N. (1993) Development and diapause induction of the Kanzawa spider mite, *Tetranychus kanzawai*. *Annual Report of the Society of Plant Protection of North Japan* 44, 184-186.

Furuhashi, K. (1990) Resurgence of citrus red mite, *Panonychus citri* McGregor induced by synthetic pyrethroids. *Japan Pesticide Information* 57, 21-26.

Gerson, U., Kenneth, R. and Muttah, T.T. (1979) *Hirsutella thompsonii*, a fungal pathogen of mites. II. Host-pathogen interactions. *Annals of Applied Biology* 91, 29-40.

Gillespie, D.R., Roitberg, B., Basalyga, E., Johnstone, M., Opit, G., Rodgers, J. and Sawyer, N. (1998) Biology and application of *Feltiella acarisuga* (Vallot) (Diptera: Cecidomyiidae) for biological control of twospotted spider mites on greenhouse vegetable crops. *Pacific Agri-food Research Centre (Agassiz) Technical Report* No. 145, 4pp.

Gimenez-Ferrer, R.M., Erb, W.A., Bishop, B.L. and Scheerens, J.C. (1994) Host-pest relationships between the twospotted spider mite (Acari: Tetranychidae) and strawberry cultivars with differing levels of resistance. *Journal of Economic Entomology* 87, 168-175.

Gokkes, M., Shecner, Z., Baum, Y. and Birati, I. (1992) Control of the common red spider mite - *Tetranychus cinnabarinus* - in standard carnations and solid aster. *Hassadeh* 72, 1013-1014.

Golovkina, L.S. and Zvereva, Yu.F. (1991) Resistant forms of *Phytoseiulus* as a component in the system of cucumber protection under greenhouse conditions. In: *Ekologicheskie osnovy primeneniya insektoakaritsidov.* VIZR, St. Petersburg, Russia, pp. 49-55.

Goodwin, S. (1990) Seasonal abundance and control of spider mites (Tetranychidae) infesting commercial strawberries in coastal New South Wales. *Journal of the Australian Entomological Society* 29, 161-166.

Gordon, S.C. (1975) Feeding damage by *Bryobia* mites to Brussels sprouts under grass. *Plant Pathology* 24, 122.

Green, A.St.J. and Dybas, R.A. (1984) Avermectin B1: control of mites on ornamentals. British Crop Protection Conference. Pests and diseases. *Proceedings of a conference held at Brighton Metropole, England, November 19-22, 1984, Volume 3.* British Crop Protection Council, Croydon, UK, pp. 1129-1133.

Gu, Y., Zhang, Y.C. and Zhang, Z.F. (1996) Study on the predation of *Stethorus punctillum* Weise to *Tetranychus urticae* (Koch). *Natural Enemies of Insects* 18, 163-166.

Hazan, A., Gerson, U. and Tahori, A.S. (1973) Life history and life tables of the carmine spider mite. *Acarologia* 15, 414-440.

Helle, W. and Sabelis, M.W. (1985a) *Spider Mites: Their Biology Natural Enemies and Control.* Volume 1A. Elsevier, Amsterdam. 405 pp.

Helle, W. and Sabelis, M.W. (1985b) *Spider Mites: Their Biology, Natural Enemies and Control. Volume 1B.* Elsevier, Amsterdam. 458 pp.

Herron, G., Edge, V. and Rophail, J. (1993) Clofentezine and hexythiazox resistance in *Tetranychus urticae* Koch in Australia. *Experimental & Applied Acarology* 17, 433-440.

Hessein, N.A. (1975) Morphology and biology of the carmine mite *Tetranychus cinnabarinus* (Boisduval) (Acarina: Tetranychidae). *Libyan Journal of Agriculture* 4, 117-122.

Ho, C.C. (1990) A preliminary study on the biological control of *Tetranychus kanzawai* in tea field by *Amblyseius fallacis* and *Phytoseiulus persimilis* (Acarina: Tetranychidae, Phytoseiidae). *Journal of Agricultural Research of China* 39, 133-140.

Ho, C.C. and Chen, W.H. (1998) Life history, food consumption, and seasonal occurrence of *Feltiella minuta* (Diptera: Cecidomyiidae) on eggplant. *Chinese Journal of Entomology* 18, 27-37.

Ho, C.C. and Chen, W.H. (2001) Evaluation of feeding and ovipositing responses of *Scolothrips indicus* (Thysanoptera: Aeolothripidae) to amounts of Kanzawa spider mite eggs (Acari: Tetranychidae). *Plant Protection Bulletin Taipei* 43, 165-172.

Ho, C.C. and Chen, W.H. (2002) Evaluation of feeding and ovipositing responses of *Oligota flavicornis* (Coleoptera: Staphylinidae) to amounts of Kanzawa spider mite eggs (Acari: Tetranychidae). *Plant Protection Bulletin Taipei* 44, 15-20.

Hoy, M.A. and Conley, J. (1987) Selection for abamectin resistance in *Tetranychus urticae* and *T. pacificus* (Acari: Tetranychidae). *Journal of Economic Entomology*

80, 221-225.

Huang, J.-H. (1994) Effects of temperature on the development, oviposition and longevity of the citrus red mite, *Panonychus citri* (McGregor). *Wuyi Science Journal* 11, 26-36.

Hussey, N.W. (1985) Cucumbers. In: Hussey, N.W. and Scope, N.E.A. *Biological Pest Control. The Glasshouse Experience*. Blandford Press, Dorset, pp. 175-179.

Hussey, N.W. and Huffaker, L.B. (1976) Spider mites. In: Delucchi, V.L. (ed.) *Studies in Biological Control*. Cambridge University Press, Cambridge, pp. 179-228.

Hussey, N.W., Read, W.H. and Hesling, J.J. (1969) *The Pests of Protected Cultivation: The Biology and Control of Glasshouse and Mushroom Pests*. Edward Arnold (Publishers) Ltd., London. 404 pp.

Jacobson, R.J., Croft, P. and Fenlon, J. (1999) Response to fenbutatin oxide in populations of *Tetranychus urticae* Koch (Acari: Tetranychidae) in UK protected crops. *Crop Protection* 18, 47-52.

Jensen, A. and Mingochi, D.S. (1988) Chemical control of red spider mite (*Tetranychus urticae* Koch) on tomatoes in Zambia. *Acta Horticulturae* 218, 275-280.

Jeppson, L.R., Baker, E.W. and Keifer, H.H. (1975) *Mites Injurious to Economic Plants*. University of California Press, Berkeley, California. 614 pp.

Jindra, Z., Taborsky, V. and Skoda, P. (1991) Spontaneous occurrence of a predatory bug *Orius majusculus* (Reut.) in glasshouses. *Ochrana Rostlin* 27, 207-209.

Karaca, I., Ghavami, M.D. and Senal, D. (2002) Effect of different host plants on the development of predatory insect species *Deraeocoris pallens* Reuther (Heteroptera: Miridae). *Acta Horticulturae* 579, 469-472.

Karg, W. (1985) A new species of spider mite of the genus *Bryobia* Koch on cucumber crops under glass. *Archiv fur Phytopathologie und Pflanzenschutz* 21, 327-330.

Kazak, C., Colkesen, T., Karut, K. and Sekeroglu, E. (1997) Biological control of *Tetranychus cinnabarinus* by *Phytoseiulus persimilis* on greenhouse cultivars. *Bulletin OILB SROP* 20(4), 215-220.

Kazak, C., Karut, K. and Sekeroglu, E. (2000) The population dynamics and predation of Hatay strain of *Phytoseiulus persimilis* Athias-Henriot (Acari: Phytoseiidae) on the prey *Tetranychus cinnabarinus* Boisduval (Acari: Tetranychidae); effects of different initial prey and predator ratios on greenhouse cucumbers. *IOBC/WPRS Bulletin* 23(1), 195-200.

Khanjani, J., Kamali, K. and Sahragard, A. (1999) Functional response of *Anystis baccarum* (L.) (Acari: Anystidae) to different densities of two spotted spider mite, *Tetranychus turkestani* U.& N. (Acari: Tetranychidae). *Agricultural Sciences and Technology* 13, 141-147.

Kielkiewicz, M. (1996) Dispersal of *Tetranychus cinnabarinus* on various tomato cultivars. *Entomologia Experimentalis et Applicata* 80, 254-257.

Kim, Y.H., Kim, J.H. and Han, M.W. (1999) A preliminary study on the biological control of *Tetranychus kanzawai* Kishida in Angelica utilis Makino by *Phytoseiulus persimilis* Anthias-Henriot (Acarina: Tetranychidae, phytoseiidae). *Korean Journal of Applied Entomology* 38, 151-155.

Kim, D.I. and Lee, S.C. (1996) Functional response and suppression of prey population of *Amblyseius womersleyi* Schicha (Acarina: Phytoseiidae) to *Tetranychus kanzawai* Kishida (Acarina: Tetranychidae). *Korean Journal of Applied Entomology* 35, 126-131.

Kolodochka, L.A. (1985a) Feeding by the predacious phytoseiid mite *Amblyseius herbarius* (Wainstein) (Parasitiformes, Phytoseiidae) on the mite *Bryobia lagodechiana* (Trombidiformes, Bryobiidae). *Vestnik Zoologii* 1985(5), 51.

Kolodochka, L.A. (1985b) Possibility of feeding by the phytoseiid mite *Phytoseiulus persimilis* (Parasitiformes, Phytoseiidae) on the bryobiid mite *Bryobia lagodechiana* (Trombidiformes, Bryobiidae). *Vestnik Zoologii* 1985(2), 54.

Koveos, D.S. and Veerman, A. (1996) Involvement of the circadian system in photoperiodic termination of diapause in the spider mite Tetranychus urticae. *Journal of Insect Physiology* 42, 681-691.

Kropezynska, D. and Tomczyk, A. (1996) Development of *Tetranychus urticae* Koch and *Tetranychus cinnabarinus* Boisd., populations on sweet pepper and *Phytoseiulus persimilis* (A.-H) effectiveness in their control. *Bulletin OILB SROP* 19(1), 71-74.

Kumar, V. and Sharma, D.D. (1993) Bio-ecology and chemical control of spider mite, *Tetranychus ludeni* Zecher on okra. *Indian Journal of Plant Protection* 21(1), 68-71.

Kuang, H.-Y. and Cheng, L.-S. (1990) Studies on differentiation between two sibling species *Tetranychus cinnabarinus* and *T. urticae*. *Acta Entomologica Sinica* 33, 109-115 [in Chinese].

Laing, J.E. (1969) Life history and life table of *Tetranychus urticae* Koch. *Acarologia* 11, 32-42.

Lindquist, E.E. (1985) External anatomy. In: Helle, W. and Sabelis, M.W. (eds) *Spider Mites: Their Biology, Natural Enemies and Control.* Volume A. Elsevier, Amsterdam, pp. 3–28.

Lindquist, R.K., Casey, M.L., Bauerle, W.L. and Short, T.L. (1987) Effects of an overhead misting system on thrips populations and spider mite-predator interactions on greenhouse cucumber. *Bulletin SROP* 10, 97-100.

Liu, T. and Sun, J.J. (1998) Interspecific competition between *Tetranychus urticae* Koch and *T. cinnabarinus* Boisduval. *Plant Protection* 24(4), 6-9.

Luo, Z.Y., Gan, G.P., Zhang, Y.L., Chen, C.L., Yao, D.M. and Xu, G.G. (1988) Population dynamics of citrus red mite *Panonychus citri* McG. and its predatory mite *Amblyseius eharai* Amitai & Swirski and the influence of pesticide applications in a citrus orchard on Changxing Island. *Contributions from Shanghai Institute of Entomology* 8, 33-42.

Makundi, R.H. and Kashenge, S. (2002) Comparative efficacy of neem, *Azadirachta indica,* extract formulations and the synthetic acaricide, Amitraz (Mitac), against the two spotted spider mites, *Tetranychus urticae* (Acari: Tetranychidae), on tomatoes, *Lycopersicum esculentum*. *Zeitschrift fur Pflanzenkrankheiten und Pflanzenschutz* 109, 57-63.

Mansour, F., Bernstein, E. and Abo-Moch, F. (1995) The potential of spiders of different taxa and a predacious mite to feed on the carmine spider mite - a laboratory study. *Phytoparasitica* 23, 217-221.

McGregor, E.A. (1943) A new spider mite on citrus in southern California (Acarina: Tetranychidae). *Proceedings of the Entomological Society of Washington* 45, 127-128.

Meyer, M.K.P. (Smith) (1981) Mites pests of crops in Southern Africa. *Science Bulletin Department of Agriculture and Fisheries Republic of South Africa* 397, 1-92.

Meyer, M.K.P. (Smith) (1987) African Tetranychidae (Acari: Prostigmata) - with reference to the world genera. *Republic of South Africa Department of Agriculture and Water Supply Entomology Memoir* 69, 1-175.

Mizell, R.F. and Schiffhauer, D.E. (1991) Kairomone response, pesticide tolerance and field efficacy of the predatory mite, *Neoseiulus collegae* (De Leon). *Journal of Environmental Horticulture* 9, 155-159.

Morros, C.M.E. and Aponte, L.O. (1994) Biology and life table of *Tetranychus ludeni* Zacher on black bean *Phaseolus vulgaris* L. *Agronomia Tropical Maracay* 44, 667-677.

Nakagawa, T. (1991) Effect of temperature and humidity on the ability of *Amblyseius longispinosus* (Evans) to regulate the population of the Kanzawa spider mite, *Tetranychus kanzawai* Kishida. *Proceedings of the Association for Plant Protection of Kyushu* 37, 201-203.

Nihoul, P. (1993) Controlling glasshouse climate influences the interaction between tomato glandular trichome, spider mite and predatory mite. *Crop Protection* 12, 443-447.

Northcraft, P.D. and Watson, T.F. (1987) Developmental biology of *Tetranychus cinnabarinus* (Boisduval) under three temperature regimes. *Southwestern Entomologist* 12, 45-50.

Osborne, L.S., Pena, J.E. and Oi, D.H. (1995) Predation by *Tapinoma melanocephalum* (Hymenoptera: Formicidae) on twospotted spider mites (Acari: Tetranychidae) in Florida greenhouses. *Florida Entomologist* 78, 565-570.

Papaioannu-Soulioti, P. (1991) *Tetranychus urticae* (Koch) control in a rosebush glasshouse by using the Stirrup-M pheromone. *Bulletin SROP* 14(5), 140-145.

Park, C.G., Lee, M.H., Yoo, J.K., Lee, J.O. and Choi, B.R. (1995) Relative toxicity of abamectin to the predatory mite *Amblyseius womersleyi* Schicha (Acari: Phytoseiidae) and twospotted spider mite *Tetranychus urticae* Koch (Acari: Tetranychidae). *Korean Journal of Applied Entomology* 34, 360-367.

Penman, D.R. and Cone, W.W. (1972) Behaviour of male twospotted spider mites in response to quiescent female deutonymphs and to web. *Annals of the Entomological Society of America* 65, 1289-1293.

Popov, S.Ya. (1988) Phytophagous mites in the greenhouse. *Zashchita Rastenii* (Moskva) 1988(1), 46-48.

Pratt, P.D. and Croft, B.A. (1998) *Panonychus citri* (Acari: Tetranychidae) on ornamental Skimmia in Oregon, with assessment of predation by native phytoseiid mites. *Pan Pacific Entomologist* 74, 163-168.

Pritchard, A.E. and Baker, E.W. (1955) A revision of the spider mite family Tetranychidae. *Pacific Coast Entomological Society Memoir Series* 2, 1–472.

Ramos, M. and Santos, R. (2002) Release of *Phytoseiulus macropilis* (Banks) (Acari: Phytoseiidae) to *Panonychus citri* (McGregor) (Acari: Tetranychidae) control in the commercial citrus nursery located in Isla de la Juventud. *Revista de Proteccion Vegetal* 15, 156-159.

Rao, P.P., Praslicka, J. and Sutakova, G. (1996) Effect of temperature and rearing method on development and fecundity of *Tetranychus urticae* (Acarina, Tetranychidae). *Biologia-Bratislava* 51, 509-516.

Richter, P. and Schulze, H.U. (1990) Investigations on the resistance spectrum of a greenhouse population of the two-spotted spider mite, *Tetranychus urticae* Koch from Gross Gaglow (district of Cottbus). *Archiv fur Phytopathologie und Pflanzenschutz* 26, 551-556.

Rossi, E. and Conti, B. (1997) Acaricide and insecticide resistance in some strains of *Tetranychus urticae* Koch (Acarina: Tetranychidae) and of its predator *Phytoseiulus persimilis* Athias-Henriot (Acarina: Phytoseiidae). *Frustula Entomologica* 20, 168-177.

Rott, A.S. and Ponsonby, D.J. (2000a) The effects of temperature, relative humidity and host plant on the behaviour of *Stethorus punctillum* Weise (Coleoptera: Coccinellidae) as a predator of the twospotted spider mite, *Tetranychus urticae* Koch (Acari: Tetranychidae). *Biocontrol* 45, 155-164.

Rott, A.S. and Ponsonby, D.J. (2000b) Improving the control of *Tetranychus urticae* on edible glasshouse crops using a specialist coccinellid (*Stethorus punctillum* Weise) and a generalist mite (*Amblyseius californicus* McGregor) as biocontrol agents. *Biocontrol Science and Technology* 10, 487-498.

Sabelis, M.W. (1981) *Biological Control of Two-spotted Spider Mites Using Phytoseiid Predators. Agricultural Research Reports 910.* PUDOC, Wageningen. 228 pp.

Sacco, M., Pasini, C., D'Aquila, F., Fadelli, P. and Tommasini, M.G. (2002) First experiences in Italy of IPM on ornamental cut foliage: *Danae racemosa* and *Fatsia japonica*. *Bulletin OILB SROP* 23, 3-8.

Sampson, A.C., King, V.J. and Lenteren, J.C. van (1996) *Macrolophus caliginosus*, field establishment and pest control effect in protected tomatoes. *Bulletin OILB SROP* 19(1), 143-146.

Sanderson, J.P. and Zhang, Z.-Q. (1995) Dispersion, sampling and potential for integrated control of twospotted spider mites (Acari: Tetranychidae) on greenhouse roses. *Journal of Economic Entomology* 88, 343-351.

Schausberger, P. and Walzer, A. (2001) Combined versus single species release of predaceous mites: predator-predator interactions and pest suppression. *Biological Control* 20, 269-278.

Shih, C.-I., Poe, S.L. and Cromroy, H.L. (1976) Biology, life table and intrinsic rate of increase of *Tetranychus urticae*. *Annals of the Entomological Society of America* 69, 362-364.

Silva, C.A.D. da, Cordao Sobrinho, F.P. and Silva, A.D. da (1999) Biologia e exigencias termicas do acaro vermelho *Tetranychus ludeni* Zacher (Acari: Tetranychidae) em algodoeiro. In: *Anais II Congresso Brasileiro de Algodao: O algodao no seculo XX, perspectivas para o seculo XXI, Ribeirao Preto, SP, Brasil, 5-10 Setembro 1999.* Empresa Brasileira de Pesquisa Agropecuaia, Embrapa Algodao, Campina Grande, Brazil, pp. 329-331.

Smith, V.A., Orr, D.B. and Hart, E.R. (1993) Economic analysis of two-spotted spider mite management on greenhouse-grown poplars. *Tree Planters' Notes* 44(4), 154-156.

Sterk, G. and Meesters, P. (1997) IPM on strawberries in glasshouses and plastic tunnels in Belgium, new possibilities. *Acta Horticulturae* 439, 905-911.

Sumangala, K. and Haq, M.A. (1994) Microhabitat maintenance by Tetranychus ludeni (Acari: Tetranychidae). *Journal of Ecotoxicology and Environmental Monitoring* 4, 3-4.

Szwejda, J. (1993) Injury symptoms and control of two spider mite species: *Tetranychus urticae* and *T. cinnabarinus* occurring on cucumbers and tomatoes. *Materialy Sesji Instytutu Ochrony Roslin* 33, 128-135

Szwejda, J. (1999) Biological efficacy of tank mixtures of acaricides and fungicides applied for controlling spider mites (*Tetranychus* spp.) on greenhouse crops of cucumber and tomato. *Vegetable Crops Research Bulletin* 51, 69-75.

Takafuji, A. and Inoue, M. (1993) The effect of permethrin spray on the dispersal of *Tetranychus kanzawai* Kishida and the abundance of its predators. *Proceedings of the Kansai Plant Protection Society* 35, 19-25.

Tian, M.-Y. and Pang, X.-F. (1997) Effects of temperature, humidity and photoperiod on the development of citrus red spider mite in laboratory. *Natural Enemies of Insects* 19, 6-10.

Tsai, S.M., Kung, K.S. and Shih, C.I. (1989) The effect of temperature on life history and population parameters of Kanzawa spider mite, *Tetranychus kanzawai* Kishida (Acarina: Tetranychidae), on tea. *Plant Protection Bulletin, Taiwan* 31, 119-130.

Valunj, A.R., Pawar, S.A., Kadam, U.K. and Khaire, V.M. (1999) A new acaricide AC-303-630 (chlorfenapyr 10 EC) against carnation mite, *Tetranychus cinnabarinus* Boisd. *Pest Management in Horticultural Ecosystems* 5, 21-23.

Vas Nunnes, M. (1986) Some aspects of induction and termination of diapause in a green stain of the mite *Tetranychus cinnabarinus* (Boisduval) Boudreaux, 1956 (Acari: Tetrancyhidae). *Experimental & Applied Acarology* 2, 315-321.

Veire, M. van de and Degheele, D. (1992) Twospotted spider mite, *Tetranychus urticae*, control with fenpyroximate and its possible use in IPM in glasshouse tomatoes and cucumbers. *Mededelingen van de Faculteit-Landbouwwetenschappen, Rijksuniversiteit Gent* 57(3A), 925-929.

Vierbergen, G. (1990) The spider mites of The Netherlands and their economic significance (Acarina: Tetranychidae). *Proceedings of the Experimental & Applied Entomology, Netherlands Entomological Society, Amsterdam* 1,158-164.

Waite, G.K. (2001) Managing spider mites in field-grown strawberries using *Phytoseiulus persimilis* and the 'pest-in-first' technique. In: Halliday, R.B., Walter, D.E., Proctor, H.C., Norton, R.A. and Colloff, M.J. (eds) *Acarology:*

Proceedings of the 10th International Congress. CSIRO Publishing, Collingwood, Australia, pp. 381-386.

Walzer, A. and Schausberger, P. (2000) Combined use of two predatory mite species for biological/integrated control of spider mites. *Forderungsdienst* 48(2), 50-51, 54-56.

Wei, X.-K., Chiu, S.-F. and Huang, Z.-X. (1989) Studies on the control of the citrus red mite *Panonychus citri* (McG.) with chinaberry seed oil. *Journal of South China Agricultural University* 10, 48-55.

Wilkin, D.R. and Warner, J.L. (1985) Control of mites invading buildings. *International Pest Control* 27, 8-9, 16.

Wilson, L.J., Bauer, L.R. and Walter, G.H. (1996) 'Phytophagous' thrips are facultative predators of twospotted spider mites (Acari: Tetranychidae) on cotton in Australia. *Bulletin of Entomological Research* 86, 297-305.

Witul, A. (1992) Life history parameters of *Tetranychus cinnabarinus* on glasshouse plants. *OEPP/EPPO Bulletin* 22, 521-528.

Witul, A. and Kielkiewicz, M. (1998) Bionomics of the carmine spider mite (*Tetranychus cinnabarinus* Boisduval) on various host plants. *Progress in Plant Protection* 38, 406-410.

Wu, Q.H. (1990) Mating behaviour of *Tetranychus cinnabarinus* (Boisduval). *Entomological Knowledge* 27, 344-346.

Wu, Q.H. and Jing, Z.Q. (1993) Study on the induction of diapause in the carmine spider mite, *Tetranychus cinnabarinus* (Boisduval). *Entomological Knowledge* 30, 335-337.

Wu, T.K. and Wu, T.K. (1995) Integrated control of *Phyllocnistis citrella, Panonychus citri,* and *Phyllocoptruta oleivora* with periodic releases of *Mallada basalis* and pesticide applications. *Chinese Journal of Entomology* 15, 113-123.

Yamada, K. and Tsutsumi, T. (1990) Injurious biology and control of Kanzawa spider mite, *Tetranychus kanzawai* Kisida, in Japanese persimmon. *Proceedings of the Association for Plant Protection of Kyushu* 36, 186-189.

Yang, Z.Q., Cao, H.G. and Chen, F.Y. (1991) A preliminary study on *Tetranychus kanzawai. Acta Agriculturae Universitatis Jiangxiensis* 13, 129-133.

Yue, B.-S. and Tsai, J.-H. (1995) *Agistemus exsertus* Gonzalez (Acari: Stigmaeidae) as a predator of citrus red mite (*Panonychus citri* [McGregor]). *Journal of the New York Entomological Society* 103, 107-113.

Zhang, Y.-X., Zhang, Z.-Q., Chen, C.-P., Lin, J.-Z. and Chen, X. (2001) *Amblyseius cucumeris* (Acari: Phytoseiidae) as a biocontrol agent against *Panonychus citri* (Acari: Tetranychidae) on citrus in China. *Systematic & Applied Acarology* 6, 35-44.

Zhang, Z.-Q. and Jacobson, R.J. (2000) Using adult female morphological characters for differentiating *Tetrancyhus urticae* complex (Acari: Tetranychidae) from greenhouse tomato crops in UK. *Systematic & Applied Acarology* 4, 69-76.

Zhang, Z.-Q. and Sanderson, J.P. (1990) Relative toxicity of abamectin to the predatory mite *Phytoseiulus persimilis* (Acari: Phytoseiidae) and twospotted spider mite (Acari: Tetranychidae). *Journal of Economic Entomology* 83, 1783-1790.

Zhang, Z.-Q. and Sanderson, J.P. (1995) Twospotted spider mites (Acari: Tetranychidae) and *Phytoseiulus persimilis* (Acari: Phytoseiidae) on greenhouse roses: spatial distribution and predator efficacy. *Journal of Economic Entomology* 88, 352-357.

Zhou, C.A., Zou, J.J., Peng, J.C., Ouyang, Z.Y., Hu, L.C., Yang, Z.L. and Wang, X.B. (1991) Predation of major natural enemies on *Panonychus citri* and its comprehensive evaluation in citrus orchards in Hunan, China. *Acta Phytophylacica Sinica* 18, 225-229.

Zilberminc, I.V., Cindadze, K.V. and Vartapetev, S.G. (1978) Entomopathogenic fungus *Entomophthora adjarica* Cinz. Vart., perspective for use to control spider mite in glasshouses. In: Begljarov, G.A. and Gekmenev, S.Ju. (eds) *Biologiceskij Metod Borby s Vrediteljami i Boleznjami Rastenij v Zakryyom Grunte*, Kolos, Moscow, pp. 125-134.

False Spider Mites 5

5.1 Introduction

The false spider mites, as the Tenuipalpidae are commonly known, also belong to the superfamily Tetranychoidea in the order Prostigmata, as does the spider mite family Tetranychidae. They are not true spider mites because they do not produce silk webbing on plants. They are also known as flat mites because most species are dorsoventrally flattened.

False spider mites are phytophagous. They are slow-moving and are usually found on the lower surface of the leaves near the midrib or veins. Some species feed on the bark. Others live in flower heads, under leaf sheaths or in galls. Only a small number of species belonging to a few genera have become pests of economic plants and they are most commonly found on tropical fruit crops and ornamental plants.

The family Tenuipalpidae has some 800 described species belonging to over 25 genera, but many others are yet to be discovered and described. The family consists of three subfamilies, with most species belonging to the Brevipalpinae and Tenuipalpinae. *Tenuipalpus* and *Brevipalpus* are the two largest genera and also the economically most important ones.

References & suggested further reading. Jeppson *et al.* (1975); Ghai and Shenhmar (1984); Sepasgosarian (1990); Smiley *et al.* (1996).

5.2 Morphological characters

False spider mites are small mites. Most species range between 200 μm and 400 μm in body length and many are orange to red in colour when alive.

False spider mites have an elongate stylophore fused by cheliceral bases. Like spider mites, the cheliceral movable digits are long, styliform and recurved proximally. As the name suggests, the palps are simple and often reduced in size (Fig. 2.1). They lack the very strong tibial claw in spider mites, and are comprised of one to five palpal segments.

The adults usually have a flattened idiosoma which is ornamented dorsally with ridges and reticulation (Figs 5.1-5.3). There is an obvious sejugal furrow separating the propodosoma and hysterosoma (Fig. 2.1). There are two pairs of eyes and three pairs of setae on the dorsal propodosoma, which often has a bifid anteromedian lobe (Fig. 5.2).

The legs are short and wrinkled and some gall-forming genera have only three pairs. The legs have five free segments and terminate in paired claws and a pad-like empodium. Both claws and the empodium have tenent hairs.

References & suggested further reading. Jeppson *et al.* (1975); Ghai and Shenhmar (1984); Smiley *et al.* (1996).

5.3 Life history and biology

Relatively little is known about the biology of false spider mites. Current knowledge is based mostly on studies of a few species.

The life cycle of the false spider mite is similar to that of spider mites and consists of the egg, larva, protonymph, deutonymph and adult stages. The eggs are ovoid (cf: spherical in spider mites). The developmental time differs among mite species and food plants, but is longer than that in spider mites. In general, it decreases with increasing temperature and is completed in about three weeks at 25°C.

Parthenogenesis is common in this family and there is often a very high female to male sex ratio.

Females of false spider mites may live as long as two months, but their reproductive rates (often one egg per day) are much lower than those of spider mites, with fecundity rarely reaching 50 offspring per female.

False spider mites overwinter as females, which hide in protected places on the undersurface of leaves, near the base of plants or under crevices of host plants. In greenhouses, they can reproduce throughout the year.

False spider mites are very slow-moving mites. Their movement and dispersal behaviour has not been studied. Long distance dispersal is most likely passive via air currents and the movement or transportation of infected plants or other objects.

References & suggested further reading. Jeppson *et al.* (1975); Trindade and Chiavegato (1994).

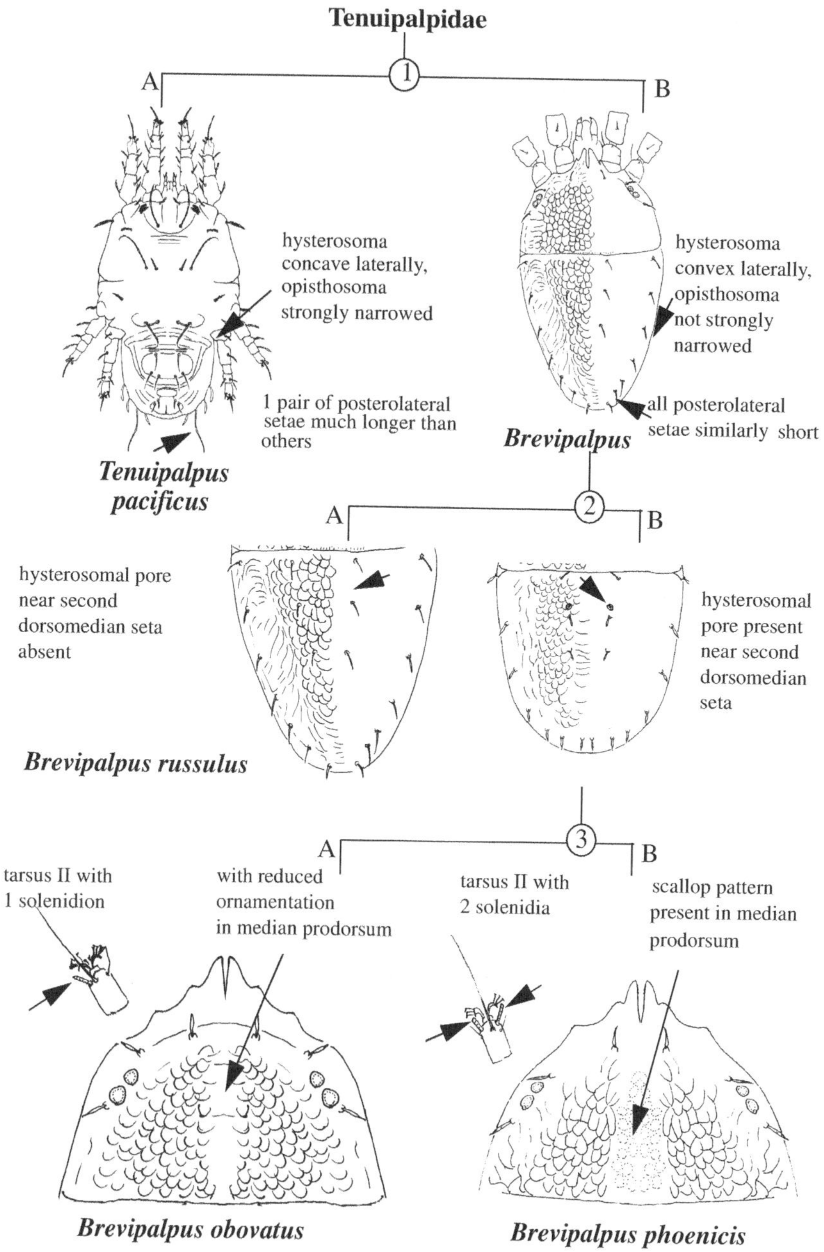

Fig. 5.1 Illustrated key to the genera and species of greenhouse false spider mites. *Tenuipalpus pacificus* redrawn and modified from Baker (1945) and *Brevipalpus russulus* redrawn and modified from Jeppson *et al.* (1975).

5.4 Species important in greenhouses

Most false spider mites attack outdoor plants and only a few of them occur in greenhouses. Four species belonging to two common genera (*Tenuipalpus* and *Brevipalpus*) are known to attack greenhouse plants. *Brevipalpus obovatus* Donnadieu and *B. phoenicis* (Geijskes) attack over 50 genera of ornamental plants. *Brevipalpus russulus* (Boisduval) mainly attacks cacti and *Tenuipalpus pacificus* Baker attacks orchids. These four species can be separated using the above illustrated key (Fig. 5.1).

5.4.1 *Brevipalpus obovatus* Dannadieu

5.4.1.1 Common Name

This species is commonly known as the privet mite because of its main host, privet. It is also known as the ornamental flat mite.

5.4.1.2 Distribution And Host Plants

This species is primarily a pest of privet and citrus. It also attacks over 50 genera of ornamental plants. It is a minor pest of greenhouse ornamental plants such as *Rhododendron, Campanula, Cissus, Gardenia* and *Hedera*. In Europe, it occurs in Austria, England, France, Spain, Cyprus and The Netherlands. In Asia and Australasia, it is distributed in Iran, Sri Lanka, Japan, Australia and New Zealand. In the Americas, it occurs in the USA, Canada, Venezuela and Argentina. In Africa, it is found in Egypt, Israel, Libya, Uganda, Kenya, Malawi, Mozambique, Angola, Zimbabwe and South Africa.

References & suggested further reading. Jeppson *et al.* (1975); Meyer (1979, 1981); Alford (1994).

5.4.1.3 Appearance And Damage Symptoms

The eggs are elliptical and bright reddish orange when first laid, but they become darker later and assume an opaque, whitish appearance just before hatching. Six-legged larvae and eight-legged nymphs are orange-red with darkish areas on the dorsal surface of the body. The colour of adult females varies from light orange to dark red with various patterns of dark pigmentation, and is correlated with host plants and amount of feeding.

This species feeds on the ventral side of leaves and also on stems and petioles. The degree of damage to plants caused by this mite varies with host species. On *Fuchsias* spp., for example, mite feeding kills cells or removes the cell contents and causes faint brown flecks on leaves. Continued mite feeding causes the flecks to merge and leaves become brownish. The upper leaf surface may become reddish as a result and the leaf may drop. On ivy, leaves are often 'cupped' and reduced in size when damaged, and growth from infested buds is usually weak and pallid.

References & suggested further reading. Jeppson *et al.* (1975); Meyer (1981).

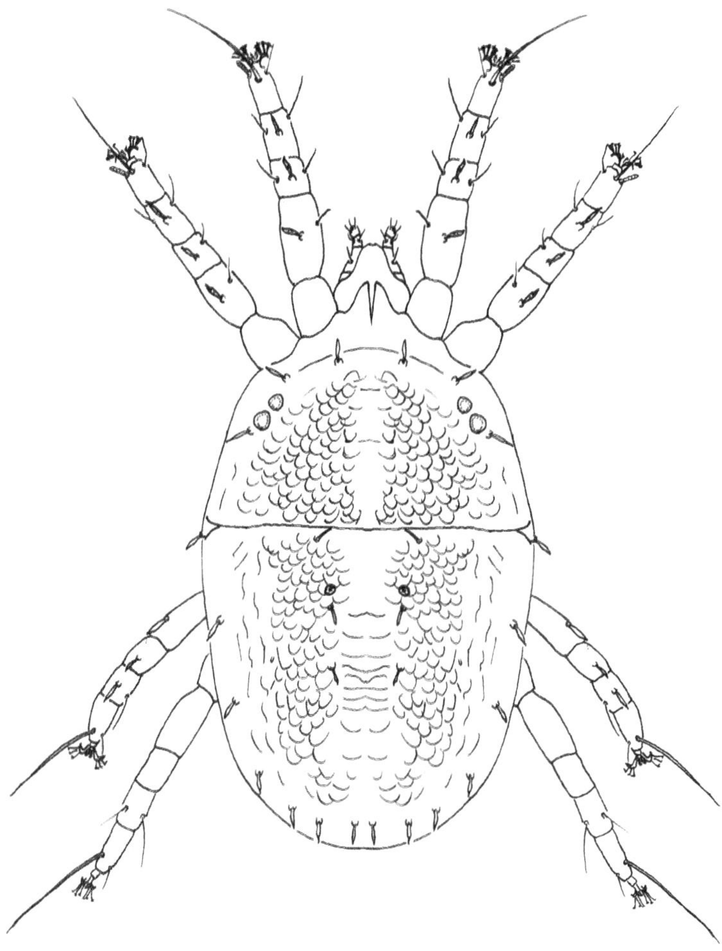

Fig. 5.2 *Brevipalpus obovatus.* Habitus of adult female, dorsal view (based in part on Meyer, 1981).

5.4.1.4 Diagnostic Characters

The adult female is flat and oval, 250-300 μm long. The idiosoma is ornamented with scalloped pattern laterally (Fig. 5.2). Cuticular striae are reduced dorsomedially on the propodosoma (cf: scallop pattern in *B. phoenicis*; Fig. 5.1.3A) and irregular dorsomedially on the hysterosoma. There are five pairs of dorsolateral setae and a pair of humeral setae on the dorsal hysterosoma. A pair of hysterosomal pores is present (Fig. 5.1.2B). The palps are four-segmented. The palptarsus has three setae. There is a single solenidion on tarsus II (cf: two in *B. phoenicis*; Fig. 5.1.3).

References and suggested further reading. Jeppson *et al.* (1975), Meyer (1979, 1981).

5.4.1.5 Life History And Biology

The eggs of this species are deposited in clusters near the midrib on the underside of the leaf. Eggs hatch in two to three weeks and immature stages last two to four weeks depending on temperature and host plants. On azalea (*Rhododendron*), development from egg to adult takes three to four weeks at 23-27°C. On peppermint, *B. obovatus* eggs have an average incubation period of eight days, whereas the immature stages last 14 days and the adult life span is 45-50 days. Mortality is low and fecundity is high at 25°C, which is about the optimal temperature for this species.

This mite breeds continuously in greenhouses under favourable conditions. Reproduction is parthenogenetic and males are rarely found.

References & suggested further reading. Jeppson *et al.* (1975); Shereef *et al.* (1984); Goyal *et al.* (1985); Trindade and Chiavegato (1994).

5.4.1.6 Control And Management

Biological control of this species in greenhouses is not possible at the moment due to the lack of effective agents, although several species of predatory mites have been tested. The phytoseiid mite *Neoseiulus idaeus* Denmark & Muma is known to be associated with this species in the field, but laboratory tests of the suitability of *B. obovatus* as prey for the phytoseiid mite show that the predator can develop on a diet of *B. obovatus* but can not reproduce on it. However, another phytoseiid predator, *Euseius scutalis* (Athias-Henriot), feeds voraciously on the adults of *B. obovatus* and can consume five adults per day. This species has been mass-produced for controlling citrus whitefly immatures in the field. The species may be mass-produced for biological control of *B. obovatus* in greenhouses.

Two other predatory mites, *Agistemus exsertus* Gonzalez (Stigmaeidae) and *Cheletogenes ornatus* (Canestrini & Fanzago) (Cheyletidae), are dependent on *B. obovatus* as a major prey on citrus in Egypt. In the field, the predator and prey populations are inversely correlated. Therefore, these two species have potential to be biological control agents of *B. obovatus*.

Currently, only some chemicals provide effective control of this species, which is susceptible to sulphur but, except diazinon, not to organophosphorus insecticides. Clorobenzilate and ovex are effective, but dicofol is most effective. Recent trials on citrus in the USA show that pyridaben and fenbutatin-oxide also give at least 35 days control of this mite.

Whenever possible, use common sense pest control methods first and chemicals as the last option (Table 1.5).

References & suggested further reading. Morishita (1954); Empson (1961); Jeppson *et al.* (1975); Heungens (1986); Childers (1994); Donia *et al.* (1995); Rezk and Gadelhak (1996); Tamai *et al.* (1997).

5.4.2 *Brevipalpus phoenicis* (Geijskes)

5.4.2.1 Common Name

The most widely accepted common name is the red and black flat mite. It is also known as the citrus leprosis mite and scarlet mite. In South Africa, it is known as the reddish black flat mite. In Australia, it is called the 'bunch mite' because it damages grape bunches.

5.4.2.2 Distribution And Host Plants

This species is mostly a pest of citrus and tea. It also attacks over 50 genera of ornamental plants, but only occasionally becomes a minor pest in greenhouses. Sometimes it occurs with *B. obovatus* on the same plant. It is a species of worldwide distribution.

5.4.2.3 Appearance And Damage Symptoms

The eggs are elliptical, reddish orange when first laid, but they become darker as development proceeds. Six-legged larvae and eight-legged nymphs are orange red with a pair of distinct lateral darkish areas on the idiosomal dorsum that become darker in the adult females, which are dark red (cf: *B. obovatus* may have darkish areas on the back, but they are not well-defined if present).

Damage symptoms vary on different host plants. Mite feeding on the under surface of leaves can produce a brownish, scurfy discoloration extending along either side of the midrib to the base of the leaf. Leaves of

Hedera appear 'cupped' after being fed upon and the tips of the lobes become necrotic.

References & suggested further reading. Jeppson *et al.* (1975); Meyer (1981).

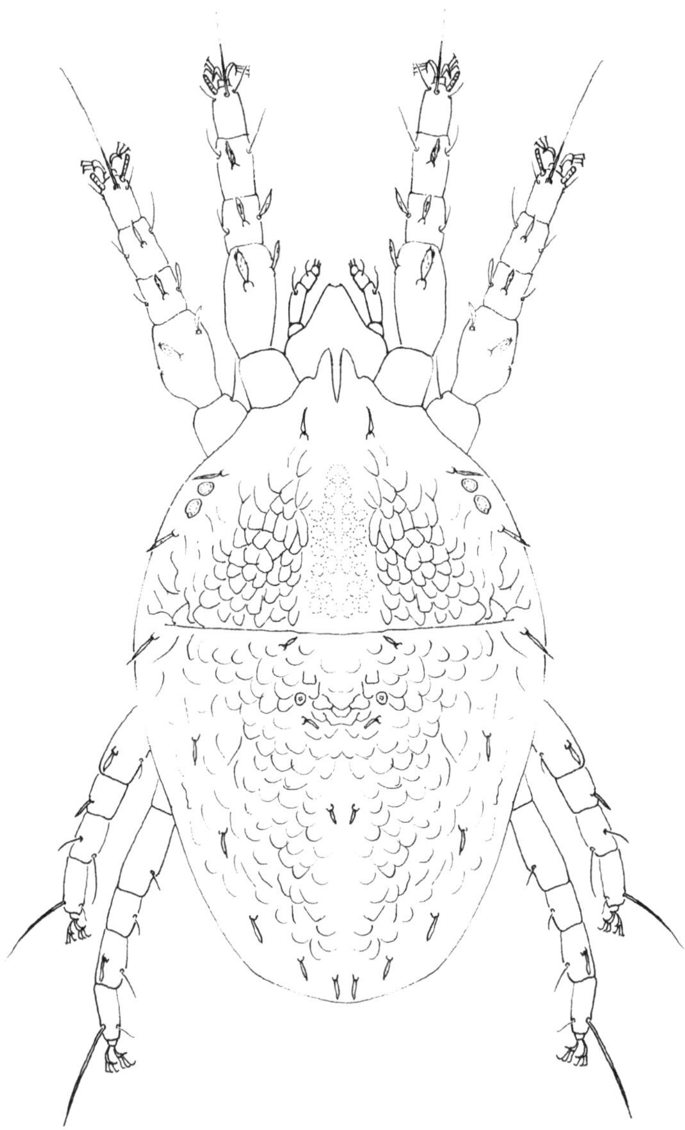

Fig. 5.3 *Brevipalpus phoenicis*. Habitus of adult female, dorsal view (based in part on Meyer, 1981).

5.4.2.4 DIAGNOSTIC CHARACTERS

The adult female of this mite has five pairs of short dorsolateral setae and one pair of humeral setae on the dorsal hysterosoma (Fig. 5.3). Like *B. obovatus*, a pair of hysterosomal pores is present. The dorsocentral surface of the propodosoma is covered with a scallop pattern (cf: reduced striae in this area in *B. obovatus*; Fig. 5.1.3), whereas the dorsocentral part of the hysterosoma has only irregular striae. There are two solenidia on tarsus II (cf: one pair in *B. obovatus*; Fig. 5.1.3).

References & suggested further reading. Jeppson *et al.* (1975); Meyer (1979).

5.4.2.5 LIFE HISTORY AND BIOLOGY

The life history of this species is generally very similar to that of *B. obovatus*. For example, on *Rhododendron* sp., development of this species from egg to adult takes 23.8-29.9 days at 23-27°C, which is not significantly different from that of *B. obovatus*. The fecundity of this species is greatly influenced by temperature and also to some degree by the host plant. More eggs are produced during summer (27 eggs) than winter (16 eggs) and during day than night.

References and suggested further reading. Zacher *et al.* (1970); Trindade and Chiavegato (1994).

5.4.2.6 CONTROL AND MANAGEMENT

Biological control of this species in greenhouses is not possible at the moment due to the lack of effective agents. A phytoseiid mite collected from orange in Brazil, *Euseius citrifolius* Denmark & Muma, has been tested in the laboratory as a predator of *B. phoenicis*. All mobile stages of the predator feed on *B. phoenicis* and prefer the immature stages (especially larvae). An increase in the predator/prey ratio resulted in higher predation levels. In the field, prey mortality rates are as high as 70-100% when the predator to prey ratio is 1:5. This species has potential to be a biological control agent of *B. phoenicis*, as do the three species that are effective against *B. ovatus* (see 5.4.1.6).

Chemicals effective against *B. obovatus* (see 5.4.1.6) are applicable to *B. phoenicis*. Common sense pest control methods should be used whenever appropriate (Table 1.5).

References & suggested further reading. Jeppson *et al.* (1975); Oomen (1982); Gravena *et al.* (1994).

5.4.3 Other species

5.4.3.1 *Brevipalpus russulus* (Boisduval)

This species has been reported from some 15 species of plants. It is a pest of cacti and succulents in Japan and several countries in Europe, North America and South America. In New Zealand, it attacks several species of ornamental plants.

Mite feeding causes the cacti to become uniformly reddish grey. Damaged plants show reduced growth.

Adult females of this mite have six pairs of short dorsolateral setae and a pair of humeral setae on the hysterosoma (Fig. 5.1.1B). The dorsal propodosoma is covered entirely with reticulation. The reticulated pattern on the posterior hysterosoma is somewhat elongate transversely. The hysterosomal pores are absent (cf: present in *B. obovatus* and *B. phoenicis*; Fig. 5.1.2). The palptarsus has three setae. There is only one solenidion on tarsus II.

Nothing is known about the biology and control of this mite, but control methods for *B. obovatus* and *B. phoenicis* should be applicable to it. Common sense pest control methods should be used whenever appropriate (Table 1.5).

References & suggested further reading. Evans *et al.* (1961); Jeppson *et al.* (1975); Ashley and Manson (1987).

5.4.3.2 *Tenuipalpus pacificus* Baker

This species is a pest of orchids in many parts of the world, especially in the Pacific areas. In Europe (especially in the UK, Germany and The Netherlands), it is sometimes reported to cause damage on greenhouse-grown orchids.

Damaged plants have dark spots on the leaves and pits appear on the upper leaf surface. Serious injury may result in necrosis of the tissue.

This species has a relatively long life cycle. The egg stage lasts 18-23 days and the larva, protonymph and deutonymph stage each takes 14-15 days.

Tenuipalpus mites differ from the ovoid species of *Brevipalpus* by the abrupt narrowing of the opisthosoma, which makes the lateral outline appear concave near the level of leg IV. Adult females of *T. pacificus* have a simple dorsal ornamentation consisting of a few longitudinal striae. The first two pairs of propodosomal setae are about as long as the posterior marginal setae on the hysterosoma but about half as long as the third pair of the dorsal propodosoma. The inner pair of medioventral setae on the propodosoma are much shorter than the outer pair, but longer on the metapodosoma (Fig. 5.1.1A). There are two pairs of genital setae. The palp is three-segmented.

Little is known about the biology and control of this species, but methods effective against other *Brevipalpus* should be applicable to it. Common sense pest control methods should be used whenever appropriate (Table 1.5).

References & suggested further reading. Baker (1945); Dosse (1954); Jeppson *et al.* (1975); Alford (1994).

5.5 References cited

Alford, D.V. (1994) *A Colour Atlas of Pests of Ornamental Trees, Shrubs and Flowers.* Wolfe Publishing Ltd., London. 448 pp.

Ashley, E. and Manson, D.C.M. (1987) The occurrence of *Brevipalpus essigi* and *B. russulus* (Acari: Tenuipalpidae) in New Zealand. *New Zealand Entomologist* 10, 131-133.

Baker, E.W. (1945) Mites of the genus *Tenuipalpus* (Acarina: Trichadenidae). *Proceedings of the Entomological Society of Washington* 47, 33-38.

Childers, C.C. (1994) Feeding injury to 'Robinson' tangerine leaves by *Brevipalpus* mites (Acari: Tenuipalpidae) in Florida and evaluation of chemical control on citrus. *Florida Entomologist* 77, 265-271.

Donia, A.R.A., Helal, E.M., El-Hamid, M.M.A. and Zakzouk, E.A. (1995) Mass rearing and field evaluation of released predaceous mite, *Euseius scutalis* (Athias-Henriot) on the citrus whitefly, *Aleurotrachelus citri* (Priesner & Hosny). *Alexandria Journal of Agricultural Research* 40, 209-219.

Dosse, G. (1954) *Tenuipalpus orchidarum* Parfitt num auch in deutschen Gewachschausern. *Zeitchrift für angewandte Entomologie* 36, 304.

Empson, D.W. (1961) *Brevipalpus obovatus* Donn. in Staffordshire. *Plant Pathology* 10, 126.

Evans, G.O., Sheals, J.G. and Macfarlane, D. (1961) *The Terrestrial Acari of the British Isles. An Introduction to Their Morphology, Biology and Classification.* British Museum, London. 219 pp.

Ghai, S. and Shenhmar, M. (1984) A review of the world fauna of Tenuipalpidae (Acarina: Tetranychoidea). *Oriental Insects* 18, 99-172.

Goyal, M., Sadana, G.L. and Sharma, N.K. (1985) Influence of temperature on the development of *Brevipalpus obovatus* (Acarina: Tenuipalpidae). *Entomon* 10, 125-129.

Gravena, S., Benetoli, I., Moreira, P.H.R. and Yamamoto, P.T. (1994) *Euseius citrifolius* Denmark & Muma predation on citrus leprosis mite *Brevipalpus phoenicis* (Geijskes) (Acari: Phytoseiidae: Tenuipalpidae). *Anais da Sociedade Entomologica do Brasil* 23, 209-218.

Jeppson, L.R., Baker, E.W. and Keifer, H.H. (1975) *Mites Injurious to Economic Plants.* University of California Press, Berkeley, California. 614 pp.

Heungens, A. (1986) Soft-skinned mites in azalea culture and comparable control results on other host plants. *Verbondsnieuws voor de Belgische Sierteelt* 30, 257-269.

Meyer, M.K.P. (Smith) (1979) The Tenuipalpidae (Acari) of Africa with keys to the world fauna. *Entomology Memoir, Department of Agriculture & Water Supply, Republic of South Africa* 50, 1-135.

Meyer, M.K.P. (Smith) (1981) Mite pests of crops in Southern Africa. *Science Bulletin Department of Agriculture and Fisheries Republic of South Africa* 397, 1-92.

Morishita, F.S. (1954) Biology and control of *Brevipalpus inornatus* (Banks). *Journal of Economic Entomology* 47, 449-456.

Rezk, H.A. and Gadelhak, G.G. (1996) Relationship between phytophagous and predatory mites in citrus orchards and the effect of acaricides on their populations. *Alexandria Journal of Agricultural Research* 41, 217-224.

Oomen, P.A. (1982) Studies on population dynamics of the scarlet mite, *Brevipalpus phoenicis*, a pest of tea in Indonesia. *Mededelingen van de Landbouwhogeschool te Wageningen* 82, 1-88

Sepasgosarian, H. (1990) Addendum to a list of the world genera and species of the family Tenuipalpidae (Acari). *Entomologischer Mitteilungen aus dem Zoologischen Museum Hamburg* 10, 85-90.

Shereef, G.M., Hanna, M.A. and Megali, M.K. (1984) Biological studies on two mite species injurious to camphor, rose and peppermint; with first description of their prelarvae. *Bulletin de la Societe Entomologique d'Egypte* 63, 49-55

Smiley, R.L., Frost, W.E. and Gerson, U. (1996) A revision of the subfamily Tegopalpinae, with the description of two new genera and five new species (Acari: Tenuipalpidae). *International Journal of Acarology* 22, 167-180.

Tamai, M.A., Moraes, G.J. de, Silva, C.A.D. da and Nunes, A.M. (1997) Suitability of *Brevipalpus obovatus* as prey to *Neoseiulus idaeus* (Acari: Tenuipalpidae, Phytoseiidae) on cassava. *Systematic & Applied Acarology* 2, 101-106.

Trindade, M.L.B. and Chiavegato, L.G. (1994) Biological characteristics of the mites *Brevipalpus obovatus* D., *B. californicus* B. and *B. phoenicis* G. (Acari: Tenuipalpidae). *Anais da Sociedade Entomologica do Brasil* 23, 189-195.

Zacher, M.A., Wafa, A.K. and Yousef, A.A. (1970) Biology of *Brevipalpus phoenicis* (Geijskes) in Egypt. *Bulletin de la Societe Entomologique d'Egypte* 54, 177-183.

Tarsonemid Mites

6

6.1 Introduction

Tarsonemid mites belong to the family Tarsonemidae of the order Prostigmata. No common names have been proposed for the family.

Many tarsonemid species are fungivores, algivores and herbivores, and others are predators of other mites, parasites of insects and possibly symbionts of insects. Some phytophagous tarsonemids are pests of agricultural crops and are important in greenhouses.

The Tarsonemidae is a large family of worldwide distribution. More than 500 species belonging to some 40 genera are known in the world and many others are yet to be discovered. The family consists of three subfamilies: Pseudotarsonemoidinae, Acarapinae and Tarsoneminae. Most of the described species are distributed in the Tarsoneminae, which includes two large genera *Tarsonemus* (over 270 species) and *Steneotarsonemus* (over 70 species). Most plant feeding species belong to a few genera in the Tarsoneminae, except *Polyphagotarsonemus*, which is a member of the Pseudotarsonemoidinae.

References & suggested further reading. Jeppson *et al.* (1975); Lindquist (1986); Lin and Zhang (2002).

6.2 Morphological characters

Tarsonemid mites are small in size (100 to 300 μm). Most species are semi-translucent, pale or whitish, but the actual colour is affected by the food ingested. Some phytophagous species assume a greenish appearance when they feed on green leaves.

Tarsonemids are dimorphic, with males differing markedly from females in both size and structure. Adult females are generally ovoid, but many species of *Steneotarsonemus* are elongate. Males are smaller than females, characterized by the subterminal genital capsule and strong leg IV with a single claw (Fig. 6.7).

The gnathosoma is capsulate (Fig. 6.4). The movable digit of the chelicerae is a fine stylet. The reduced palps are closely appressed.

Adult females have stigmata but no peritremes; the opening of the stigmata is on the anterior lateral dorsum of the propodosoma (Fig. 6.2.4). Females have a pair of capitate trichobothria and two pairs of simple setae on dorsal propodosoma (Fig. 6.3), whereas males have three or four pairs of simple setae.

The dorsal surface of the opisthosoma is covered with a series of plates (Fig. 6.3). The structure, length and location of dorsal setae on the plates are used in classification. Coxal apodemes are strongly developed.

Leg IV of females are three-segmented and slender, terminating in a characteristic long seta (Fig. 6.4). Leg IV of males are generally four-segmented and strong (Fig. 6.7), terminating in a claw (in some species the tarsus and tibia of leg IV are fused as in Fig. 6.1.1). Legs II and III usually terminate in a pair of claws and a membranous empodium.

References & suggested further reading. Jeppson *et al.* (1975); Lindquist (1986).

6.3 Life history and biology

The life cycle of tarsonemid mites consists of egg, larva and adult stages, but there is a quiescent nymph inside the larval cuticle. Eggs are often laid singly, but some species lay eggs in small clusters. Larvae are active and feed like adults, with a few exceptions: larvae of *Acarapis* do not move due to reduced legs II and III but feed normally, whereas the larvae of *Iponemus* are capable of movement but move little and do not feed.

With the loss of active nymphs, the life cycle is very short, usually less than a week at ambient temperatures for most species. The developmental time of each species varies with host species, temperature and other environmental conditions and is often completed within a week.

Like spider mites, unfertilized eggs of tarsonemids give rise to males only and fertilized eggs mostly to females. However, there are a few exceptions: unfertilized females of some populations of *P. pallidus* produce offspring of both sexes, whereas some populations of *P. pallidus*, *T. confusus* Ewing and *T. fusarii* Cooreman are known to be thelytokous.

Females lay one to five eggs per day for one to two weeks, with fecundity being ten to 20 eggs in most species. Females live considerably longer than males. Sex ratio varies with local conditions among species and is strongly female-biased in many species.

Males of tarsonemids are known to find pharate females still contained in the larval cuticle and carry the female for as long as 24 hours — a behaviour known as 'precopulation'. True copulation ensues as soon as female adults emerge.

There have been no studies or reports on the diapause in Tarsonemidae so far. Greenhouse tarsonemids reproduce throughout the year.

Tarsonemids disperse as adult females. Parasitic species disperse with their hosts. Other species can cover short distances by walking and long distances via phoretic association with flying insects.

Feeding habits of this family are unknown for most species. Many species are fungivorous. Species of *Phytonemus, Polyphagotarsonemus* and *Steneotarsonemus* are mostly phytophagous and are of economical importance. A few species of *Hemitarsonemus* and *Tarsonemus* also attack plants.

References & suggested further reading. Schaarschmidt (1959); Jeppson *et al.* (1975); Lindquist (1986).

6.4 Species important in greenhouses

Most phytophagous tarsonemids attack outdoor plants and a few of them have been reported from plants in greenhouses. *Polyphagotarsonemus latus* is undoubtedly the most important tarsonemid mite and has been known as an important pest of many crops and ornamentals in greenhouses wordwide. It is the equivalent of the twospotted spider mite in the Tarsonemidae. *Phytonemus pallidus* is also an important pest of strawberries and many ornamental plants in greenhouses and it also occurs on a worldwide basis. A few other species are also known to occasionally occur in greenhouses and are minor pests: *Hemitarsonemus tepidariorum* (Warburton), *Steneotarsonemus laticeps* (Halbert), *Xenotarsonemus belemnitoides* Weis-Fogh, *Tarsonemus confusus* Ewing and *T. bilobatus* Suski. One species, *Tarsonemus floricolus* Canestrini & Fanzago, was recorded from decaying buds of *Iris kaempferi* in a greenhouse in Brooklyn Botanic Garden, New York, but this species is fungivorous and is known as a pest in mushroom houses. It is therefore unlikely to be a plant pest. All these species and genera can be separated in the following key (Figs 6.1 and 6.2).

References & suggested further reading. Davis (1938); Ewing (1939); Karl (1965b); Hussey *et al.* (1969); Jeppson *et al.* (1975); Nemestothy (1983); Nakao (1991); Alford (1994).

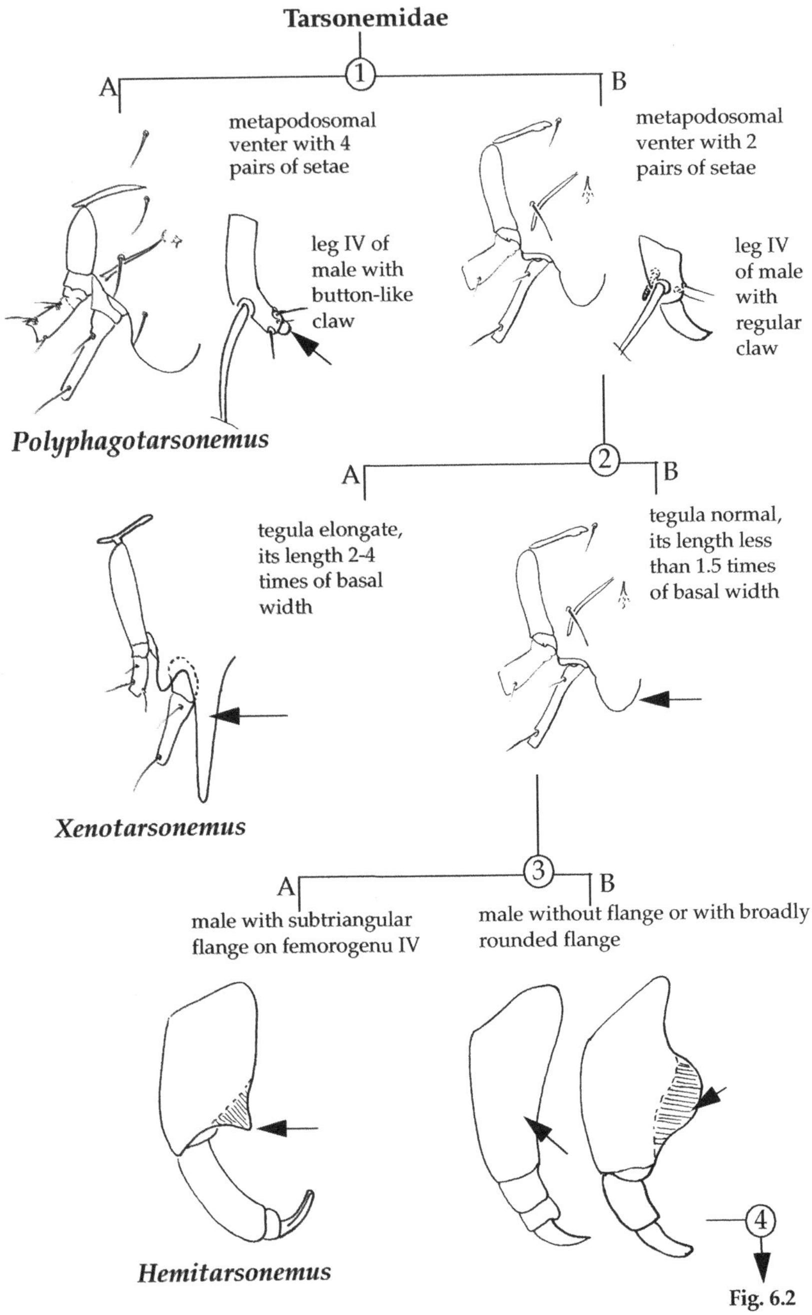

Fig. 6.1 Key to genera and species of greenhouse Tarsonemidae. Part I.

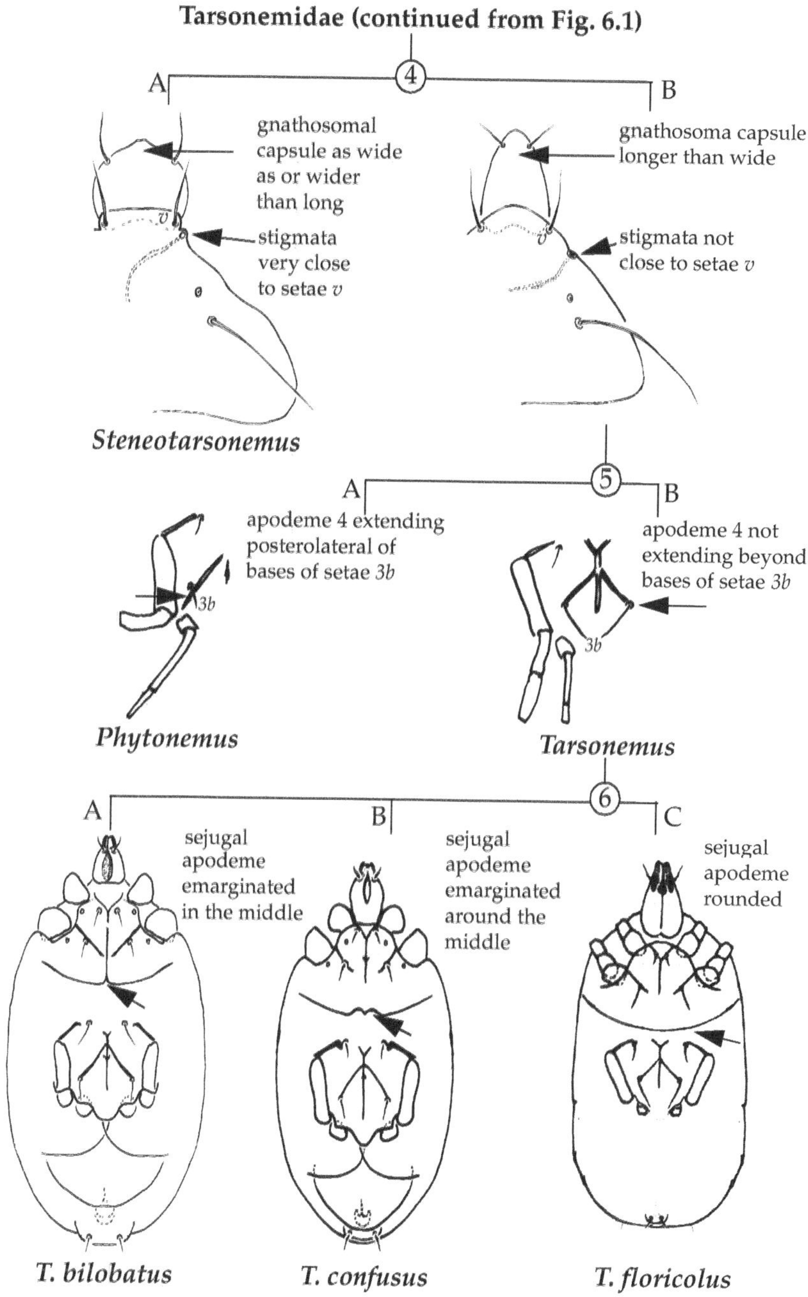

Fig. 6.2 Key to genera and species of greenhouse Tarsonemidae. Part II (some of the figures are modified from Schaarschmidt, 1959 and Kaliszewski, 1993).

6.4.1 *Polyphagotarsonemus latus* (Banks)

6.4.1.1 Common Name

This species is widely known as the broad mite. It is also known as the yellow (tea) mite, white mite, citrus silver mite or tropical mite.

6.4.1.2 Distribution And Host Plants

It is a major mite pest throughout the tropics and also in greenhouses in temperate regions. As its name suggests, *P. latus* is polyphagous; it attacks plants belonging to 60 families, among which are economic crops and ornamental plants such as pepper, tomato and cucumber, African violet, begonia, chrysanthemum, cyclamen, dahlia, *Gloxinia*, *Fuchsia*, *Gerbera*, *Hibiscus*, *Impatiens* and ivy (*Hedera*).

References & suggested further reading. Jeppson *et al.* (1975); Gerson (1992); Ciampolini *et al.* (1989); Lin and Zhang (2002).

6.4.1.3 Appearance And Damage Symptoms

The eggs are oval and have a flattened base where they attach to the leaf surface. The eggs are mostly translucent except the upper surface with longitudinal rows of white tubercles. The six-legged larvae are whitish or pale when first hatched but soon become translucent. The adult females are broadly oval and rich amber or dark green (colour depending on host plants) with an indistinct white strip in the mid-dorsum. The adult males are broadly short and tapered posteriorly. They are colourless at first but become rich amber when fully developed. Males are only half as long as females but they have relatively longer legs than females. They also lack the white dorsal stripe present in the mid-dorsum of females.

Broad mites are often found on young leaves and feed mostly on the under surface of the leaves. Damaged leaves of *Gerbera* may split or crack open and have a rugged appearance. Injured flowers have distorted and discoloured rays. There is often sudden curling and wrinkling of leaves followed by discoloration or blistering. Plant growth may stop and survival of the plant may be threatened when severely injured. Damage of cucumber, aubergines and *Solanum aviculare* includes crinkling, cracking, discoloration, malformations, swelling and necrosis similar to those caused by a hormonal weedkiller. The symptoms can persist for many weeks after the removal of mites.

References & suggested further reading. Jeppson *et al.* (1975); Meyer (1981); Roditakis and Drossos (1987); Costilla *et al.* (1994); Cho *et al.* (1996a).

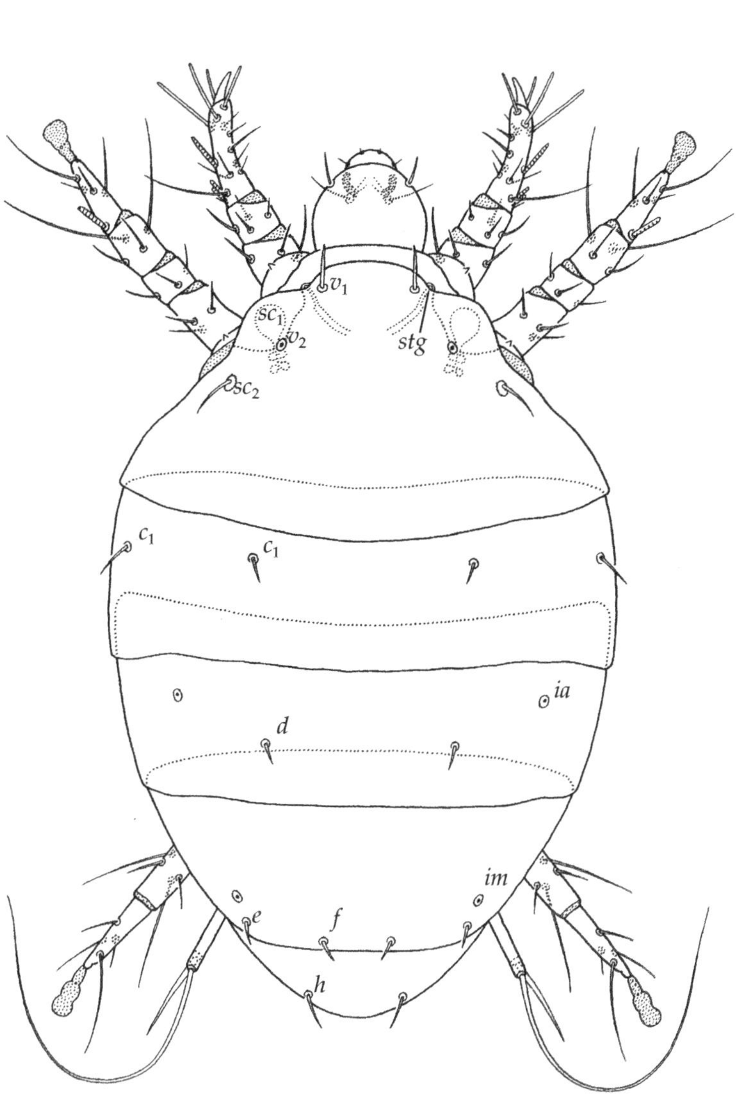

Fig. 6.3 *Polyphagotarsonemus latus*. Habitus of adult female, dorsal view with notation of structures; trichobothria sc_2 is covered by the prodorsal shield and hence shown in dotted line; stg = stigmata (drawn by D. Helmore after Lindquist, 1986).

6.4.1.4 Diagnostic Characters

Adult females of this mite are small (about 200 μm) and have an unornamented dorsal shield (Fig. 6.3). The prodorsal shield is not enlarged to cover the stigmata. The prodorsum has a pair of captitate trichobothria and two pairs of setae (Fig. 6.3). Dorsal idiosomal setae are short. The metapodosomal ventral has four setae (Fig. 6.4). Tibiotarsus I has a single strong claw. There are four pairs of setae on the prodorsum in the male. Tibia and tarsus IV of the male are fused and terminate in a button-like claw (Fig. 6.1A), which is characteristic of this species.

References & suggested further reading. Jeppson *et al.* (1975); Lindquist (1986); Cho *et al.* (1993).

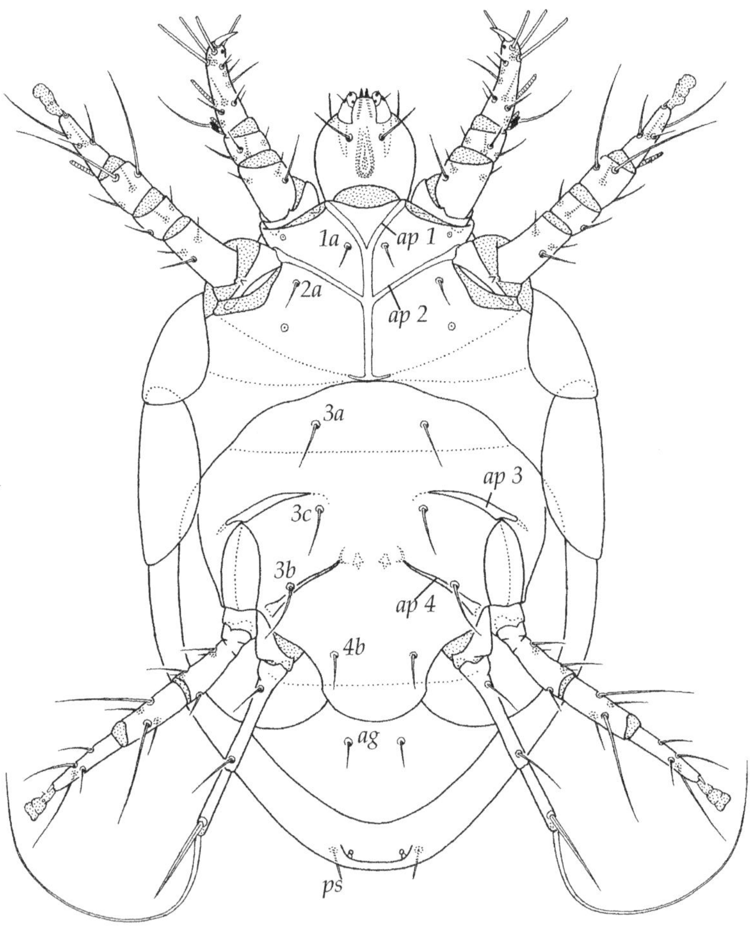

Fig. 6.4 *Polyphagotarsonemus latus*. Habitus of adult female, ventral view with notation of structures; ap 1-4 = apodemes 1-4 (drawn by D. Helmore after Lindquist, 1986).

6.4.1.5 Life History and Biology

Life history traits vary on different host plants and varieties. Developmental time decreases with temperature and is often less than a week in greenhouses (Fig. 6.5). On pepper, the developmental period from egg to adult averages 4.1 and 4.1 days at 25°C for males and females, respectively. Adult female and male longevity is 11 and 15 days, respectively. Adult females lay 25 eggs. The female/male sex ratio is 2.8 in the laboratory, and 2.3 on seedlings in the greenhouse. The intrinsic rate of increase was 0.359, the finite rate of increase 1.43 individuals/female per day, the mean generation time 10.34 days and the net reproductive rate 41.0.

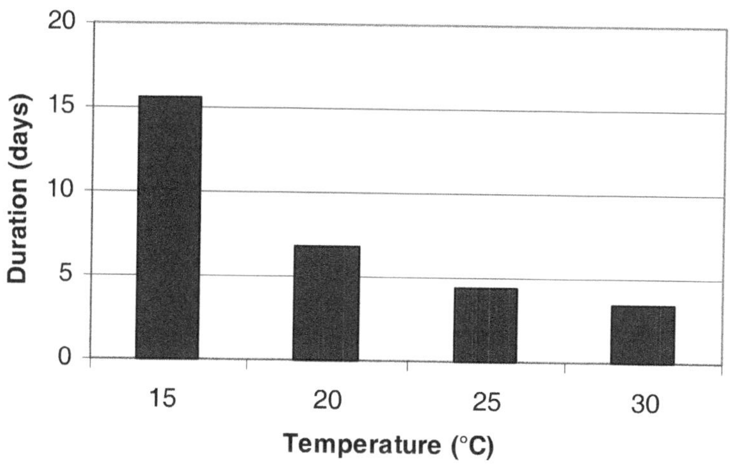

Fig. 6.5 *Polyphagotarsonemus latus*. Developmental time (egg to adult) at different temperatures on *Phaseolus vulgaris* (drawn from data in Lee *et al.*, 1992).

This species disperses by various means. Short distance movement may be accomplished through walking. Mites may reach far away uninfested plants by wind. Human transport of infested plants is another way of spreading this mite. This mite also disperses through insects living on plants. Females of *P. latus* have a phoretic relationship with the whitefly *Bemisia tabaci* on beans (*Phaseolus vulgaris*) in Colombia and on watermelons in Venezuela and most often attach to the tarsi and tibiae of *B. tabaci.* The greenhouse whitefly, *Trialeurodes vaporariorum* is also known to disperse *P. latus* in greenhouses in China. In the USA, this species is known to disperse on *Bemisia argentifolii* in greenhouses.

References & suggested further reading. Flechtmann and Rosa (1980); Aubert *et al.* (1981); Karuppuchamy and Mohanasundaram (1987); Natarajan (1988); Ho (1991); Lee *et al.* (1992); Parker and Gerson (1994); Karmakar *et al.* (1996a); Fan and Petitt (1998); Gui *et al.* (1998); Vieira and Chiavegato (1999); Wu *et al.* (2000).

6.4.1.6 Control and Management

Biological control

Several phytoseiid species are known to attack the broad mite and some of them have been demonstrated to be effective in greenhouses.

Neoseiulus barkeri, which is commercially available in many countries, can effectively reduce populations of *P. latus* from more than 100 mites per leaf to zero in a week on *Capsicum*, when released at the rate of ten or more predatory mites per plant. Three weekly releases of five predatory mites per main stem provide adequate protection of the plant from mite injury for over seven weeks. Another commercially available species, *N. cucumeris* is also effective in a similar way.

Neoseiulus californicus and *Euseius ovalis* feed and reproduce very well on *P. latus*. The latter, when released at the predator:prey ratio of 1:20, can cause a rapid decline in adult population of *P. latus*. They are promising candidates for use in greenhouses for *P. latus* control.

Neoseiulus agrestis can consume *P. latus* and reproduce on it in the laboratory but it can not reduce the abundance of *P. latus* on azalea plants in the greenhouse. Augmentative releases may work for this species.

Neoseiulus longispinosus and *Typhlodromalus peregrinus* are able to consume adults and immature stages of *P. latus* and their potential as a biocontrol agent should be investigated.

Conidia of *Beauveria bassiana* can cause 88% mortality of *P. latus* in controlled laboratory conditions and may be trialed in greenhouses.

References & suggested further reading. Kolodochka and Prutenskaya (1987); Hariyapa and Kulkarni (1988); Pena *et al.* (1989); Pena (1992); Petitt (1992); Castagnoli and Falchini (1993); Fan and Petit (1994); Karuppuchamy *et al.* (1994); Pena and Osborne (1996); Pena *et al.* (1996); Wang *et al.* (2000); Manjunatha *et al.* (2001).

Chemical control

Because biological control works well for this species, the use of chemicals should be minimized. Some chemicals may sometimes induce mite problems instead of suppressing them. An example is the use of dichlorvos, cypermethrin and fluvalinate, monocrotophos, methyl-O-demeton [demeton-O-methyl], formothion, thiometon and ethion on *Capsicum* in India. This mite may be difficult to control using non-systemic pesticides on certain plants because mites may be able to hide in curly leaves, which make full coverage of sprays difficult.

Many chemicals are effective against this species, but there is the issue of availability and registration in different countries. No general recommendations will be made here. Some examples are given below, but users are encouraged to consult local pest control workers for advice.

In India, dicofol, bromopropylate, azocyclotin and avermectin (abamectin) can eliminate the mite population in two weeks, and abamectin (at 18 g/litre, applied at 25 ml/100 litres water) remain effective for up to three weeks. Wettable sulphur is also known to be effective.

In Thailand, 0.07% prothiofos, 0.075% formetanate, 0.072% triazophos and 0.071% methiocarb provide effective control of *P. latus*.

In China, liuyangmycin [an antibiotic preparation from *Streptomyces griseolus*] applied at 25 ppm gives good control of this mite on *Capsicum*, with an efficiency equivalent to control by dicofol applied at 200 ppm.

In Korea, dicofol EC, chinomethionat WP, pyridaben WP, and pyraclofos WP give effective control of this mite on pepper. Elsewhere, bromopropylate 50% at 0.5%, hexythiazox 10% at 0.4-0.5%, dicofol 21% + tetradifon 7.5% at 1.5-2%, and endosulfan 35% at 2% are recommended for preventive control of this mite on *Capsicum*.

In Brazil, two applications (spaced seven days) of abamectin (5.4 g AI/ha) and triazophos (400 g AI/ha) give significant control for up to 28 days.

References & suggested further reading. Heungen and Degheele (1986); David (1991); Liu *et al.* (1991); Sepswasdi *et al.* (1991); Xie *et al.* (1992); Costilla *et al.* (1994); Cho *et al.* (1996b); Karmakar *et al.* (1996b); Scarpellini (1999).

Integrated control

Selective acaricides such as abamectin can be used together with biological control. Organic material such as Manipueira, a liquid extract from cassava roots, can provide effective control of *P. latus* when diluted in water (1:3) and sprayed three times at weekly intervals. Plant resistance is not yet employed as a component in broad mite control, but there is good potential. In Cuba, for example, a double haploid of sweet pepper that has higher mean fruit weight and yield is known to be tolerant of *P. latus*.

References & suggested further reading. Gerson (1992); Depestre and Gomez (1995); Ponte (1996).

6.4.2 *Phytonemus pallidus* (Banks)

6.4.2.1 Common Name

This species is most commonly known as the cyclamen mite. It is also known as the strawberry mite because it is a major pest of strawberries.

6.4.2.2 Distribution And Host Plants

It is a widespread species and is known from North America, South America, Asia, Australasia, Europe and Africa. It is a pest of strawberries and watercress, and many ornamental plants such as cyclamen, African violet, azalea, begonia, carnation, chrysanthemums, *Gerbera* and ivy.

References & suggested further reading. Eyndhoven and Groenewold (1959); Jeppson *et al*. (1975); Lin and Zhang (2002).

6.4.2.3 Appearance And Damage Symptoms

The eggs are pale, smooth, oval and relatively large (125 x 75 μm). At high densities, concentrations of eggs may appear to the naked eye as a fine coat of dust. The six-legged larvae are translucent, pearly-white, about 200 μm long and 80 μm wide. The adult females are ovoid, pale to yellowish brown. They are about twice as long as eggs. The adult males are smaller (170 μm).

This species prefers young leaves or flower buds. Eggs are often laid within unopened buds and between adpressed halves of the dorsal surfaces of folded leaves. On leaves they feed on the upper side. Infested leaves are twisted, curled, distorted, brittle and reduced in size. Damaged *Gerbera* produces bronzed patches along the midribs and slight curling of the leaf. Rays of flower are deformed when flowers are attacked in the bud stage. Damaged strawberry leaves have wrinkled upper surfaces, irregular folding and fluting of the leaf margin; veins bulge upward like blisters. Mildly damaged plants have an unnatural dense appearance because petioles fail to elongate. Severely damaged leaves become brittle, turn brown or silvery, and eventually die, and flowers and young fruits become brown near the base, and may turn black and die.

References & suggested further reading. Hussey *et al*. (1969); Jeppson *et al*. (1975); Meyer (1981).

6.4.2.4 Diagnostic Characters

The adult females are about 250 μm long. There are two pairs of setae on dorsal propodosoma; the second pair much longer than setae on dorsal hysetrosoma. The trichobothria are capitate, partly or not covered by lateral margins of the prodorsal shield (Fig. 6.6). Apodemes 4 extend beyond the bases of setae *3b*. Leg IV of male has tibia and tarsus fused. The propodosoma of male has four pairs of dorsal setae and the fourth is much shorter than the third pair and is laterad of the line forming by the first three setae.

References & suggested further reading. Jeppson *et al*. (1975); Meyer (1981); Lindquist (1986).

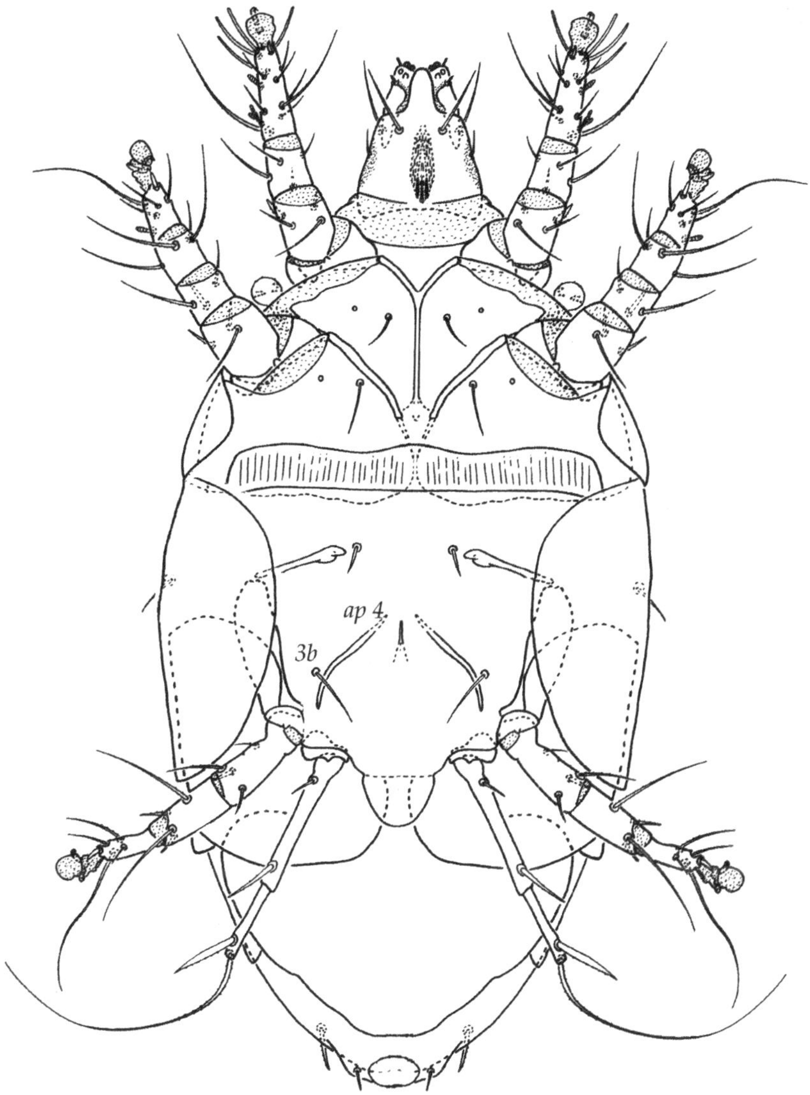

Fig. 6.6 *Phytonemus pallidus*. Habitus of adult female, ventral view with notation of structures; *ap 4* = apodeme 4 (modified from Lindquist, 1986).

6.4.2.5 LIFE HISTORY AND BIOLOGY

The life cycle is completed in one to three weeks depending on temperature and is usually completed in ten to 14 days in greenhouses. The incubation period of the eggs is three to 13 days; it is four days at 20°C but

13 days at 15°C. Larvae feed for one to four days. Adults emerge in two to seven days. Adult females lay eggs in batches. Each female lays one to three eggs per day and a total of 12-16 eggs during the oviposition period. Sex determination is complex in this species. In addition to normal haplo-diploidy, unfertilized females of some populations are known to produce both males and female offspring, whereas other populations are known to be thelytokous. Female:male sex ratio is high, varying from 2.3:1 to 5.1:1.

This species prefers dark and wet habitats. Dispersal is passive by wind, phoresy on insects and other animals, movement of infested plants or equipment. Reproduction is continuous throughout the year if temperature is favourable such as in greenhouses.

References & suggested further reading. Garman (1917); Karl (1965a); Hussey *et al.* (1969); Jeppson *et al.* (1975); Meyer (1981).

6.4.2.6 CONTROL AND MANAGEMENT

Biological control

Neoseiulus cucumeris is used in many parts of Europe to control of *P. pallidus* on strawberries and can provide effective control at lower cost than chemical control. It should be released at ten to 20 mites per plant or the predator:prey ratio of 1:10 as soon as the symptoms of mite damage are noticed. Another two phytoseiid species, *Neoseiulus reductus* in Russia and *Neoseiulus reticulatus* in the USA, are also effective for *P. pallidus* control on strawberries.

Like *N. cucumeris, N. californicus* can also significantly reduce *P. pallidus* densities on strawberries as well as providing effective control of *T. urticae* on the same crop in greenhouses.

In Northwest USA, *N. fallacis* or *N. cucumeris* are both effective predators of *P. pallidus* on strawberries, but the former performs slightly better than the latter regardless of whether *T. urticae* is present on the crop. Although *N. fallacis* can provide more rapid control of both pests, *N. cucumeris* seems to give longer-term regulation at lower densities.

References & suggested further reading. van Driesche and Hauschild (1987); Malov and Tokunova (1990); Radetskii and Polyakova (1991); Croft *et al.* (1998); Meshkov-Yu (2000); Easterbrook *et al.* (2001); Petrova *et al.* (2002); Tuovinen (2002).

Chemical control

Many chemicals are effective against this species, but there is the issue of availability and registration on different crops and in different countries. No general recommendations will be made here. Some examples are given below for information, but users are encouraged to consult local pest control workers for advice.

On strawberries, effective control can be achieved using: endrin at 0.025% or endosulfan at 0.05% at four-week intervals before the opening of the first flower; triazophos, amitraz and pyridaben with two applications at seven-day intervals after fruit harvest and leaf mowing; pirimiphos-methyl applied at 1.6 kg/ha; abamectin at 1.12 and 2.24 kg AI/ha.

On cyclamen, control can be achieved using: endrin at 0.025% or endosulfan at 0.05% or dicofol at 0.03% at four-week intervals at first sign of damage.

On *Saintpaulia*, dienochlor, triazophos, fenbutatin-oxide and oxamyl are effective when applied at first sign of damage.

On *Peperomia* and *Saintpaulia*, mites can be controlled by drenching plants in 0.05% schradan at the rate of 142 ml per pot (10.2 cm) at three- or four-week intervals.

References & suggested further reading. Hussey *et al.* (1969); Jeppson *et al.* (1975); Bashkatova *et al.* (1983); Tusnadi and Kerenyi-Nemestothy (1989); Welch (1989); Goodwin (1990); Labanowska (1992); Tuovinen (2000).

Integrated control

A variety of approaches can be used. First of all, it helps greatly to plant healthy plants, with good hygiene in plant production. Strawberry planting material can be kept free of mites or completely disinfected on acquisition by immersing them in warm water at 45°C for 13-15 minutes and then in cold water. Immersion of plants in hot water at 43.5 °C for 30 minutes kills all the mites. A better alternative is to treat loosely stacked plants with saturated air at 44.5 °C for one hour. After the treatment, plants should be dried before packing and planted as soon as possible.

Immersion of cyclamen, *Crassula* and *Saintpaulia* in water at 15°C for seven minutes is effective against this mite but may also injure plants.

It is impotant to monitor the level of infestation and introduce predatory mites at the right time and rates. Biological control should be integrated with chemical control for other pests by careful and selective use of pesticides to prevent harmful effects on predatory mites.

Plant resistance has not yet been used in cyclamen mite control, but there is evidence that the potential exists. Some varieties of strawberry are known to have higher levels of resistance to cyclamen mite.

References & suggested further reading. Hussey *et al.* (1969); Jeppson *et al.* (1975); Bashkatova *et al.* (1983); Heungens (1986); Khokhryakova and Polyakova (1986); Labanowska (1992); Titov (1986); Tuovinen (2000); Hellqvist (2002).

6.4.3 Other tarsonemid species

6.4.3.1 *Hemitarsonemus tepidariorum* (Warburton)

Known as the fern mite, this species is known only from the USA and the UK. It is phytophagous and feeds on various species of ferns. In the UK, it causes injury to ferns grown in greenhouses. It mainly attacks *Asplenium*, but is also present on *Polystichum* and *Pteris*.

This species is only found on younger parts of the plant and is never on old fronds. It prefers protected areas (grooves, hollows) on the pinnae or the stalk, but is also found among the scale leaves and scales at the base of the plant. Mite feeding produces minute depressions on the frond surface, which later become brown speckles. Damaged fronds become distorted and swollen. Heavy infestation may arrest plant growth and even kill the plant.

The eggs are oval and pearly white, nearly half as long as adult females (about 230 μm). The six-legged larvae are pale, white and elongate. The adult females are oval and yellowish brown. Males are smaller than females, suboval and yellow. The fourth leg has a distally located triangular flange on the inner margin of the femorogenu and a prominent tarsal claw (Fig. 6.7).

This species prefers dark, warm and humid conditions. Eggs are laid near the tip of the fronds and hatch in a few days in greenhouses. The life cycle is about two and a half weeks. The incubation period of the eggs lasts two to four days. Larvae feed for one to two weeks and then become quiescent. Adults emerge in three to four days and start to lay eggs in a couple of days. They lay one or two eggs per day, depositing them singly or in groups, in a protected niche within distorted pinnae, on the upper surface of the leaf stalks near the tips of fronds or between young scales near the base of the plant. The adults live for four to 16 days.

This species overwinters as eggs and adults. Adult males are rarely found in the winter. They can disperse by walking over a considerable distance on damp surfaces of the greenhouse bench. They are mainly spread by the movement of infested plants.

There has been no report on the natural enemies of this species, but generalist phytoseiid predators may be tried against this species. The species is best controlled by good sanitation and prevention of infestation of plants. For other control measures, see sections on *P. latus* and *P. pallidus*.

References & suggested further reading. Cameron (1925); Hussey *et al.* (1969); Lindquist (1986); Alford (1994); Lin and Zhang (2002).

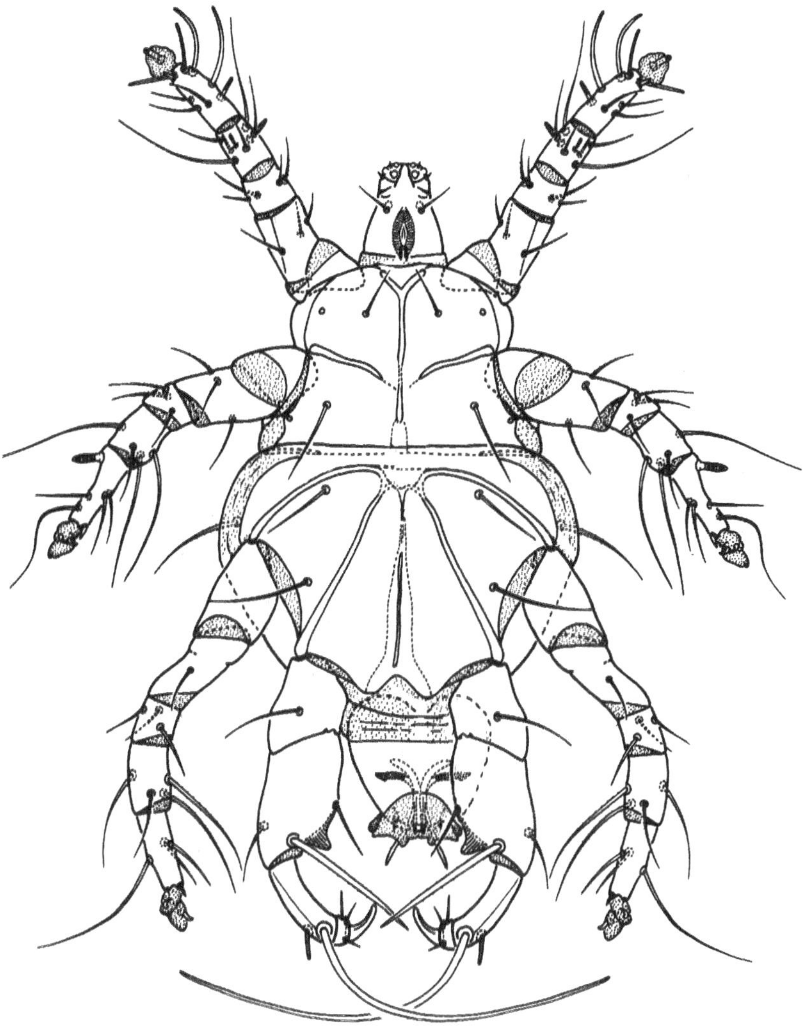

Fig. 6.7 *Hemitarsonemus tepidariorum*. Habitus of adult male, ventral view (modified from Lindquist, 1986).

6.4.3.2 *Steneotarsonemus laticeps* (Halbert)

Known as the bulb scale mite, this species is distributed in Europe (Ireland, the UK, The Netherlands, Sweden, Poland, Ukraine), in South Africa and on the West Coast of the USA. This species is a pest of *Amaryllis*, forced *Narcissus* and *Hippeastrum*. It also attacks *Eucharis*, Scarborough lily, *Sprekelia* and other members of the Amaryllidaceae.

Infested bulbs of *Amaryllis* have spotted vegetative growth and red scars. The flowers from infested bulbs are weak and may be malformed. Infested bulbs of *Narcissus* produce distorted, sickle-shaped leaves and small, malformed flowers. Mite feeding removes the surface wax from the leaves so that they are often bright green instead of greyish. Later, infected leaves often develop yellow scars and saw-like edges. Heavy infestation may result in lower yield and even the death of the bulbs. Transverse cracks may also form on leaves and stems as a result of mite feeding. Stored bulbs are abnormally dry when infested, and display longitudinal brown streaks of dead tissue in the neck region when sliced across about 6 mm below the apex.

The eggs are oval and translucent. The six-legged larvae are colourless. The adult females are oval and colourless initially, but turn pale pink as they develop and assume a bright semi-translucent appearance. The male looks similar to the female in appearance but are smaller.

The adult females are about 200 μm long. The anteromedian apodeme extends just slightly beyond extremities of apodemes 2, but far away from the sejugal apodeme (Fig. 6.8). The adult males are about 160 μm long. The femorogenua IV lack a flange. The third pair of prodorsal setae are about two-thirds as long as the width of the gnathosoma and the other prodorsal setae are in a linear arrangement.

Development from egg to adult is shortest at 20°C (15 days) on bulb scales of narcissus in the laboratory, but it may take seven weeks in the field. The adults lay most eggs at 20°C, averaging 1.5 per day with a fecundity of up to 30 eggs per female. The greatest percentage of the eggs hatch at 15°C (92.8%). This species prefers warm, humid conditions.

No predatory mites have been known to feed on these mites. The best control is prevention. Bulbs with suspected symptoms should be checked and infested ones should be disposed of. Bulbs may be disinfected prior to planting. Hot water treatment is sometimes effective against mites in bulbs. Mites are killed when narcissus bulbs are immersed for four hours at 43.3°C or three hours at 44.4°C. Partial control to reduce injury to flowers may be achieved by immersing fully dormant bulbs for one hour at 43.3°C. Forced *Narcissus* bulbs can be drenched with diluted acaricides (0.1% endrin or 0.1% endosulfan) a few days after they are in heated greenhouses.

Dipping *Hippeastrum* bulbs in a 0.5% solution of endosulfan 35 EC for three hours and in hot water (46°C) for two hours can successfully reduce the mite population without undue damage to plants. However, the size of the bulb can affect the efficacy of the immersion treatment; larger bulbs require more time than small bulbs for the same effect. Crop damage can also be significantly reduced during forcing by a drenching spray, particularly of dienochlor WP and pirimiphos-methyl 50 EC at 0.3%.

References & suggested further reading. Schaarschmidt (1959); Hussey *et al.* (1969); Jeppson *et al.* (1975); Meyer (1981); Labanowski and Jaworski (1992); Alford (1994); Lynch and Bedi (1994); Doorduin *et al.* (1997).

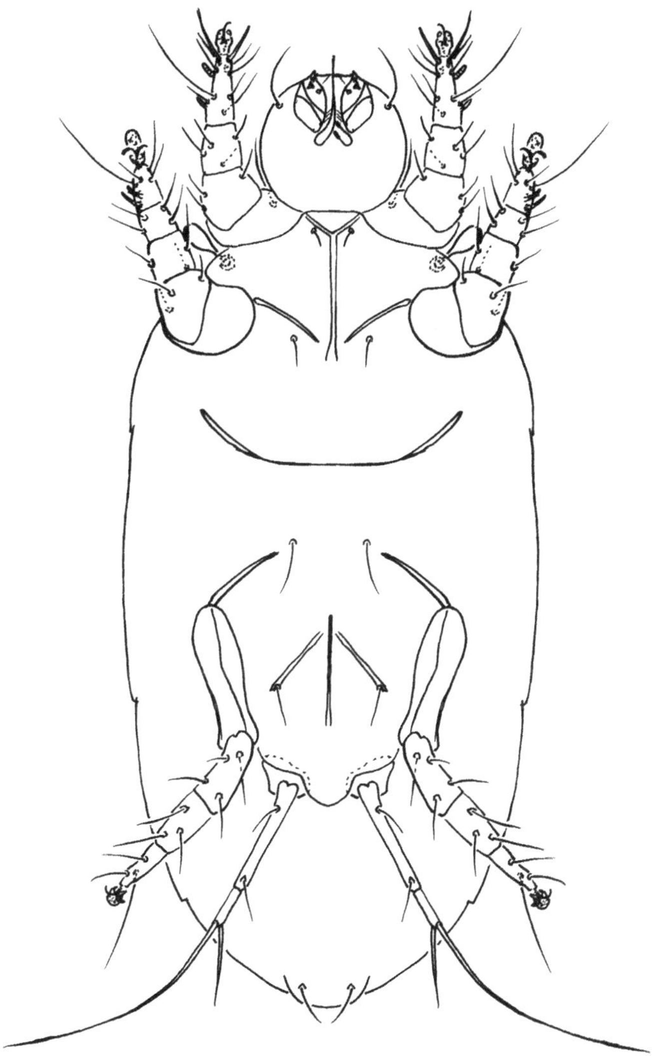

Fig. 6.8 *Steneotarsonemus laticeps* (Halbert). Habitus of adult female, ventral view (modified from Schaarschmidt, 1959).

6.4.3.3 *Xenotarsonemus belemnitoides* (Weis-Fogh)

This species is known from Denmark, Belgium, The Netherlands, Ukraine and China. It is mainly found in the soil, but also from food crops (e.g. rice and *Fragaria ananassa)* and ornamentals (e.g. *Sinningia hybrida* and *Hedera helix*, azalea). In recent years, it has become a pest of azalea grown in greenhouses in Belgium and The Netherlands.

This is a small species, with adult females measuring 170 μm in length. The anteromedian apodeme reaches beyond extremities of apodeme 2. The sejugal apodeme is broadly rounded. Apodeme 3 extends laterally beyond insertions of trochanter III. The tegula is elongate, with its length about three times of basal width (Fig. 6.1.2A). Leg IV is very short, not reaching the posterior end of the idiosoma. The tibiotarsus I is very long, terminating in a strong sickle-shaped claw. One proximal seta on tibiotarsus I is much longer than the others and is as long as the tibiotarsus.

Very little is known about the damage symptoms, life history and biology of this species. In greenhouse azalea, this species is common in the soil and moves between plants and soil. The populations of mites found on the leaves therefore continuously fluctuate. In addition, there is a high natural mortality in the population. This species is also known to co-occur with *T. confusus* in greenhouse azalea.

Sprayed at five-day intervals for four applications, methomyl at 0.35 g/litre produces reasonable mite mortality (about 75%), but dichlorvos at 0.5 g/litre water has no effect on mite populations and methamidophos at 0.4 g/litre can cause 67% mite mortality initially, but can not keep the population level down for five weeks. Abamectin, pyridaben, chlorfenapyr and chlorpyrifos used at the recommended rates give the best mite control, causing mite mortalities between 86 and 93%.

References & suggested further reading. Schaarschmidt (1959); Heungens (1993); Heungens and Tirry (2000); Lin and Zhang (2002).

6.4.3.4 *Tarsonemus confusus* Ewing

Commonly known as the confused tarsonemid mite, this species is distributed in North Amercia (USA, Canada), Europe (Turkey, Italy, Ireland, Germany, Poland, Byelorussia, Ukraine, Russia), East Asia (Japan, Korea, China) and Africa (Egypt). It has been found on many plant species, in soil and litter, in house dust and in birds' nests. It is a primarily fungivorous species and is also known as a minor pest of some ornamentals (e.g. African violet, azalea, *Cissus*, *Cyclamen*, *Gloxinia*, ivy and *Pilea*) in greenhouses in Europe and tomatoes in North America.

The adult females are 200-250 μm long. In the female, the prodorsal sensillus is elongate-capitate, 14 μm long, with numerous fine spines. The prodorsal plate has a dorsomedian apodeme. The last pair of prodorsal setae sc_2 are over three times as long as the first pair of dorsomedian setae c_1. The distance between cupules *ih* is greater than that between setae *f*. Ventrally, the anteromedian apodeme has a knot-like thickening and is not connected with apodemes 2. The sejugal apodeme is characteristically emarginated around the middle (Fig. 6.2.6B). Tibiotarsus I terminates in a sickle-shaped claw. The tibial solenidion is short and located near the base of the segment. The tarsal solenidion is located near the middle of the tibiotarsus and is as long as the width of the segment.

This species often occurs in association with *P. latus* and *P. pallidus*. Its density is usually much lower than those of the other two species. It rarely causes primary damage to host plants. It is also known to co-occur with *X. belemnitoides* in greenhouse azalea.

In greenhouse azalea, this species lives in the soil and moves between plants and soils. The populations of mites on the leaves continuously fluctuate. In addition, there is a high natural mortality in the population.

On greenhouse azalea, amitraz and pyridaben give the best mite control (mortality of 100 and 97%, respectively), followed by abamectin, endosulfan, methomyl and dicofol (mortality ranging from 76 to 94%). Treatments with bifenthrin, methamidophos and omethoate result in less than 64% mite morality.

References & suggested further reading. Schaarschmidt (1959); Hussey *et al.* (1969); Shoukry *et al.* (1990); Kaliszewski (1993); Alford (1994); Heungens and Tirry (2000); Lin and Zhang (2002).

6.4.3.5 *Tarsonemus bilobatus* Suski

This mite seems to be a widespread species and has been reported from Central America (Costa Rica) and Europe (Italy, Hungary, Poland, Byelorussia, Ukraine), Asia (Japan, Korea, China, India) and Africa (Egypt). It has been collected from many plant species, fungal and bacteria cultures, stored food and products, litter and soil. Like many other species of *Tarsonemus*, *T. bilobatus* is primarily fungivorous. It is known to cause injury to several ornamental flowers in greenhouses in Poland. It is also found on seedlings of melon, watermelon, cucumber and Chinese cabbage in greenhouses in Japan.

On seedlings of cucumber, Chinese cabbages and other plants, the symptoms are lustrous, discoloured and deformed leaves with irregular folding of the upper surface.

The eggs are laid singly on leaves. They are ovoid and translucent to pale. Adults are light brown. Females are ovoid and 200-250 µm long and males are smaller, about 170 µm long. In the female, the prodorsal sensillus is elongate-capitate, 14 µm long, with numerous fine spines. The prodorsal plate lacks a dorsomedian apodeme. The last pair of prodorsal setae sc_2 is about 2.5 times as long as the first pair of dorsomedian setae c_1. The distance between cupules *ih* is shorter than that between setae *f*. Ventrally, the anteromedian apodeme is interrupted in the anterior part, not connecting with apodeme 2, but joining posteriorly with medially emarginated sejugal apodeme (Fig. 6.2.6A).

Developmental speed varies with temperature; developmental time from egg to adult is completed in a week for females at 25°C; at the same temperature, males develop slightly faster.

Like many other *Tarsonemus*, this species seems quite polyphagous; it can be mass-reared on yeasts and have been known to contaminate fungal and bacterial cultures in the laboratory. Injury of this species to green plants in Hungary and Japan is quite unusual. More studies are needed to understand its relationship with host plants and its biology.

Phytoseiid predators effective against other tarsonemids may be applied to this species. Likewise, chemicals used against other tarsonemid mites are likely applicable to this species. As always, use common sense control methods before any use of pesticides (Table 1.5).

References & suggested further reading. Nemestothy (1983); Vargas and Ochoa (1990); Nakao (1991); Ito (1993); Kaliszewski (1993); Lin and Zhang (2002).

6.4.3.6 *Tarsonemus floricolus* Canestrini & Fanzago

This species is known from Italy, Germany, the UK, China, Japan and the USA. It mainly occurs in stored food and mushroom houses, but has been collected from grasses, gooseberry and ornamental flowers. There is one report of this species from decaying buds of *Iris kaempferi* in a greenhouse in Brooklyn Botanic Garden, New York.

Little is known about the biology and control of this species. This species seem fungivorous and is known as a pest in mushroom houses. It is unlikely to be a plant pest.

References & suggested further reading. Ewing (1939); Lin and Zhang (2002).

6.5 References cited

Alford, D.V. (1994) *A Colour Atlas of Pests of Ornamental Trees, Shrubs and Flowers*. Wolfe Publishing Ltd., London. 448 pp.

Aubert, B., Lossois, P. and Marchal, J. (1981) Mise en évidence des dégâtes causes par *Polyphagotarsonemus latus* (Banks) sur parayers à l'ile de la Réunion. *Fruits* 36, 9-24.

Bashkatova, T.V., Prokof'ev, M.A., Ogneva, L.I. and Shamanskaya, L. (1983) Tested in orchards in Siberia. *Zashchita Rasteni* 1983(8), 1-30.

Cameron, W.P.L. (1925) The fern mite (*Tarsonemus tepidariorum* Warburton). *Annals of Applied Biology* 12, 93-112.

Castagnoli, M. and Falchini, L. (1993) Suitability of *Polyphagotarsonemus latus* (Banks) (Acari, Tarsonemidae) as prey for *Amblyseius californicus* (McGregor) (Acari, Phytoseiidae). *Redia* 76, 273-279.

Cho, M.R., Chung, S.K. and Lee, W.K. (1993) Taxonomic study on cyclamen mite (*Phytonemus pallidus*) and broad mite (*Polyphagotarsonemus latus*). *Korea Journal of Applied Entomology* 32(4), 433-439.

Cho, M.R., Jeon, H.Y., Kim, D.S., Chung, B.S., Yiem, M.S. and Kim, S.B. (1996a) Host plants and damage of broad mite (*Polyphagotarsonemus latus*) on horticultural crops. *RDA Journal of Agricultural Science, Crop Protection* 38(1), 516-525.

Cho, M.R., Jeon, H.Y., La, S.Y., Kim, D.S. and Yiem, M.S. (1996b) Damage of broad mite, *Polyphagotarsonemus latus* (Banks), on pepper growth and yield and its chemical control. *Korean Journal of Applied Entomology* 35(4), 326-331.

Ciampolini, M., Maiulini, C. and Robusto, F. (1989) The mite *Polyphagotarsonemus latus* invades Italian greenhouses. *Informatore Agrario* 55, 97-101.

Costilla, M.A., Willink, E., Osores, V.M. and Venditti, M.E. (1994) The white mite *Polyphagotarsonemus latus* on pepper. *Avance Agroindustrial* 14(56), 33-34.

Croft, B.A., Pratt, P.D., Koskela, G. and Kaufman, D. (1998) Predation, reproduction, and impact of phytoseiid mites (Acari: Phytoseiidae) on cyclamen mite (Acari: Tarsonemidae) on strawberry. *Journal of Economic Entomology* 91(6), 1307-1314.

David, P.M.M. (1991) Resurgence of yellow mite *Polyphagotarsonemus latus* (Acarina: Tarsonemidae) on chilli following application of insecticides. *Madras Agricultural Journal* 78, 88-91.

Davis, A.C. (1938). *Tarsonemus* spp. attacking mushrooms. *Journal of Economic Entomology* 31, 547.

Depestre, T. and Gomez, O. (1995) New sweet pepper cultivars for Cuban off season production. *Capsicum & Eggplant Newsletter* 1995 No. 14, 47-49.

Doorduin, J.C., Lilien, K.H., Borochov, A. and Halevy, A.H. (1997) Factors influencing the duration of hot-water treatment of amaryllis bulbs (Hippeastrum). Proceedings of the seventh international symposium on flower bulbs, Herzliya, Israel. *Acta Horticulture* 2, 695-698.

Driesche, R.G. van and Vittum, P. (1987) Potential for increased use of biological control agents against greenhouse pests in Massachusetts. *Research Bulletin, Massachusetts Agricultural Experiment Station* No. 718, 88-111

Easterbrook, M.A., Fitzgerald, J.D. and Solomon, M.G. (2001) Biological control of strawberry tarsonemid mite *Phytonemus pallidus* and two spotted spider mite *Tetranychus urticae* on strawberry in the UK using species of *Neoseiulus* (*Amblyseius*) (Acari: Phytoseiidae). *Experimental and Applied Acarology* 25(1), 25-36.

Ewing, H.E. (1939) A revision of the mites of the subfamily Tarsoneminae of North America, the West Indies and the Hawaiian Islands. *Technical Bulletin of United States Department of Agriculture* 653, 1-63.

Eyndhoven, G.L. van and Groenewold, H. (1959) On the morphology of *Steneotarsonemus pallidus* and *S. fragariae* (Acar., Tars.). *Entomologische Berichten* 19, 123-124.

Fan, Y.Q. and Petitt, F.L. (1994) Biological control of broad mite, *Polyphagotarsonemus latus* (Banks), by *Neoseiulus barkeri* Hughes on pepper. *Biological Control* 4, 390-395

Fan, Y.Q. and Petitt, F.L. (1998) Dispersal of the broad mite, *Polyphagotarsonemus latus* (Acari: Tarsonemidae) on *Bemisia argentifolii* (Homoptera: Aleyrodidae). *Experimental & Applied Acarology* 22(7), 411-415.

Flechtmann, C.H.W. and Rosa, A.E. (1980) Winged beans, *Psophocarpus tetragonolobus* (L.) DC. in Sao Paulo, Brasil - preliminary observations. *Anais da Sociedade Entomologica do Brasil* 9(2), 289-292.

Garman, P. (1917) *Tarsonemus pallidus* Banks, a pest of geraniums. *Maryland Agricultural Experimental Station Bulletin* 208, 327-342.

Gerson, U. (1992) Biology and control of the broad mite, *Polyphagotarsonemus latus* (Banks) (Acari: Tarsonemidae). *Experimental & Applied Acarology* 13, 163-178.

Goodwin, S. (1990) Controlling cyclamen mite. *Australian Horticulture* 88, 56.

Gui, L.Y., Meng, G.L. and Gong, X.W. (1998) Laboratory population life table of *Polyphagotarsonemus latus*. *Plant Protection* 24(6), 10-11.

Hariyapa, A.S. and Kulkarni, K.A. (1988) Biology and feeding efficiency of the predatory mite *Amblyseius longispinosus* (Evans) on chilli mite *Polyphagotarsonemus latus* (Banks). *Journal of Biological Control* 2, 131-132

Hellqvist, S. (2002) Heat tolerance of strawberry tarsonemid mite *Phytonemus pallidus*. *Annals of Applied Biology* 141, 67-71.

Heungens, A. (1986) Soft-skinned mites in azalea culture and comparable control results on other host plants. *Verbondsnieuws voor de Belgische Sierteelt* 30, 257-269.

Heungens, A. (1993) Chemical control of the soft mite *Tarsonemoides belemnitoides* Weis Fogh (Tarsonemidae) in azalea culture. *Parasitica* 49(1-2), 3-9.

Heungens, A. and Degheele, D. (1986) Control of the broad mite, *Polyphagotarsonemus latus* (Banks), with acaricides on *Psophocarpus tetragonolobus* and *Ricinus communis*. *Parasitica* 42, 3-10.

Heungens, A. and Tirry, L. (2000) Curative chemical control of the tarsonemid mites *Tarsonemus confusus* and *Tarsonemoides belemnitoides* in azalea culture. *Parasitica* 56(4), 123-130.

Ho, C.C. (1991) Life history of *Polyphagotarsonemus latus* (Banks) feeding on lemon, tea and pepper. *Journal of Agricultural Research of China* 40, 439-444.

Hussey, N.W., Read, W.H. and Hesling, J.J. (1969) *The Pests of Protected Cultivation: The Biology and Control of Glasshouse and Mushroom Pests.* Edward Arnold (Publishers) Ltd., London. 404 pp.

Ito, Y. (1993) *Tarsonemus bilobatus* Suski. In: Ehara, S. (ed.) *Plant Mites of Japan in Colors.* Zenkoku Noson Kyoiku Kyokai, Tokyo, pp. 34-35.

Jeppson, L.R., Baker, E.W. and Keifer, H.H. (1975) *Mites Injurious to Economic Plants.* University of California Press, Berkeley, California. 614 pp.

Kaliszewski, M. (1993) Key to Palearctic of the Genus *Tarsonemus.* Acari, Tarsonemidae. Uniwersytet im. *Adama Mickiewicza w Poznantu, Seria Zoologia* Nr 14, 1-204.

Karl, E. (1965a) Untersuchungen zur morphologie und okologie von tarsonemiden gartnerischer kulturpflanzen I. *Tarsonemus pallidus* Banks. *Biologisches Zentralblatt* 84, 47-80.

Karl, E. (1965b) Untersuchungen zur morphologie und okologie von tarsonemiden gartnerischer kulturpflanzen II. *Hemitarsonemus latus* (Banks), *Tarsonemus confusus* Ewing, *T. talpae* Schaarschmidt, *T. setifer* Ewing, *T. smithi* Ewing und *Tarsonemoides belemnitoides* Weis-Fogh. *Biologisches Zentralblatt* 84, 331-357.

Karmakar, K., Sarkar, P.K., Somchoudhury, A.K. and Mukherjee, A.B. (1996a) Influence of host plants on different life stages of *Polyphagotarsonemus latus* (Banks) (Acari: Tarsonemidae). *Annals of Entomology* 14(2), 41-45.

Karmakar, K., Sarkar, P.K., Somchoudhury, A.K. and Mukherjee, A.B. (1996b) Effectiveness of some modern pesticides against different stages of yellow mite, *Polyphagotarsonemus latus* (Banks) (Acari: Tarsonemidae) infesting chilli. *Annals of Entomology* 14(2), 47-54.

Karuppuchamy, P. and Mohanasundaram, M. (1987) Bioecology and control of chilli muranai mite, *Polyphagotarsonemus latus* (Banks) (Tarsonemidae: Acari). *Indian Journal of Plant Protection* 15(1), 1-4.

Karuppuchamy, P., Balasubramanian, G., Sundarababu, P.C. and Gopalan, M. (1994) A potential predator of chilli mite, *Polyphagotarsonemus latus* (Banks) (Tarsonemidae: Acari). *Madras Agricultural Journal* 81(10), 552-553.

Khokhryakova, T.M. and Polyakova, L.T. (1986) Sources of multiple resistance to pathogens in strawberry in the Nonchernozem zone of the RSFSR. *Nauchno-Tekhnicheskii Byulleten' Vsesoyuznogo Ordena Lenina i Ordena Druzhby Narodov Nauchno-Issledovatel'skogo Instituta Rastenievodstva imeni N. I. Vavilova* No. 16, 59-62.

Kolodochka, L.A. and Prutenskaya, M.D. (1987) Ability of the phytoseiid mite *Amblyseius agrestis* to feed on the tarsonemid mite *Hemitarsonemus latus. Vestnik Zoologii* 1987(2), 80.

Labanowska, B.H. (1992) Effectiveness of acaricides in the control of strawberry mite (*Phytonemus pallidus* ssp. *fragariae* Zimm.). *Fruit Science Reports* 19(3), 137-146.

Labanowski, G. and Jaworski, A. (1992) Control of the bulb scale mite - *Steneotarsonemus laticeps* (Halbert) on *Hippeastrum. Prace Instytutu Sadownictwa i Kwiaciarstwa w Skierniewicach. Seria B, Rosliny Ozdobne* 17,179-188.

Lee, S.H., Park, C.G. and Choi, K.M. (1992) Tarsonemid mite; morphology, damage symptom, development and occurrence in Korea. *Research Reports of the Rural Development Administration, Crop Protection* 34(2), 55-62.

Lin, J.-Z. and Zhang, Z.-Q. (2002) *Tarsonemidae of the World: Key to Genera, Geographical Distribution, Systematic Catalogue & Annotated Bibliography.* Systematic & Applied Acarology Society, London. 440 pp.

Lindquist, E.E. (1986) The world genera of Tarsonemidae (Acari: Heterostigmata): morphological, phylogenetic, and systematic revision, with a reclassification of family-group taxa in the Heterostigmata. *Memoirs of the Entomological Society of Canada* 136, 1-517.

Liu, T.S., Wang, W. J., Wang, Y.S. (1991) Survey on the hosts damaged by the broad mite and its control. *Plant Protection Bulletin (Taipei)* 33, 344-353.

Lynch, S.M.T. and Bedi, A. (1994) A novel technique for culturing the bulb scale mite (*Steneotarsonemus laticeps*) and its implications for studies on biology and control. *Proceedings of Brighton Crop Protection Conference, Pests and Diseases, vol. 2. British Crop Protection Council, BCPC Publications* 1994, 583-588.

Malov, N.A. and Tokunova, M.V. (1990) Open-air rearing of *Amblyseius. Zashchita Rastenii (Moskva)* 1990(6), 22.

Manjunatha, M., Hanchinal, S.G. and Kulkarni, S.V. (2001) Interaction between *Amblyseius ovalis* and *Polyphagotarsonemus latus* and efficacy of *A. ovalis* on chilli mite and thrips. *Karnataka Journal of Agricultural Sciences* 14, 506-509.

Meshkov-Yu, I. (2000) Breeding and use of the predatory tick *Neoseiulus cucumeris* for control of the transparent dug-out tick. *Zashchita i Karantin Rastenii* 2000(6), 38-39.

Meyer, M.K.P. (Smith) (1981) Mite pests of crops in Southern Africa. *Science Bulletin Department of Agriculture and Fisheries Republic of South Africa* 397, 1-92.

Nakao, H. (1991) Studies on acarid mites (Acari: Astigmata) damaging vegetable plants. II. Damage to vegetable seedlings. *Japanese Journal of Applied Entomology and Zoology* 35, 303-309.

Natarajan, K. (1988) Transport of yellow mite *Polyphagotarsonemus latus* by cotton whitefly. *Current Science* 57, 1142-1143.

Nemestothy, K.K. (1983) The tarsonemid species occurring in Hungary (Acari: Tarsonemidae). *Novenyvedelem* 19(5), 198-202.

Parker, R. and Gerson, U. (1994) Dispersal of the broad mite, *Polyphagotarsonemus latus* (Banks) (Heterostigmata: Tarsonemidae), by the greenhouse whitefly, *Trialeurodes vaporariorum* (Westwood) (Homoptera: Aleyrodidae). *Experimental & Applied Acarology* 18(10), 581-585.

Pena, J.E. (1992) Predator-prey interactions between *Typhlodromalus peregrinus* and *Polyphagotarsonemus latus*: effects of alternative prey and other food resources. *Florida Entomologist* 75, 241-248.

Pena, J.E., Baranowski, R.M. and Denmark, H.A. (1989) Survey of predators of the broad mite in southern Florida. *Florida Entomologist* 72, 373-377.

Pena, J.E. and Osborne, L. (1996) Biological control of *Polyphagotarsonemus latus* (Acarina: Tarsonemidae) in greenhouses and field trials using introductions of predacious mites (Acarina: Phytoseiidae). *Entomophaga* 41(2), 279-285.

Pena, J.E., Osborne, L.S. and Duncan, R.E. (1996) Potential of fungi as biocontrol agents of *Polyphagotarsonemus latus* (Acari: Tarsonemidae). *Entomophaga* 41(1), 27-36.

Petrova, V., Cudare, Z. and Steinite, I. (2002) The efficiency of the predatory mite *Amblyseius cucumeris* (Acari: Phytoseiidae) as a control agent of the strawberry mite *Phytonemus pallidus* (Acari: Tarsonemidae) on field strawberry. *Acta Horticulturae* 567(2), 675-678.

Petitt, F.L. (1992) Biological control in the integrated pest management program at the Land, EPCOT Center. *Bulletin OILB SROP* 16(2), 129-132.

Ponte, J.J. da (1996) Efficiency of manipueira for the control of papaya white mite (preliminary note). *Revista de Agricultura Piracicaba* 71, 259-261.

Radetskii, V.R. and Polyakova, A.D. (1991) Phytophagous and predatory mites on strawberry. *Zashchita Rastenii* 1991(6), 14-16.

Roditakis, N.E. and Drossos, N.I. (1987) First record of *Polyphagotarsonemus latus* (Banks 1904) (Acari: Tarsonemidae) on greenhouse peppers in Crete. *Entomologia Hellenica* 5, 35-36.

Scarpellini, J.R. (1999) Evaluation of acaricides against the white mite *Polyphagotarsonemus latus* (Banks, 1904) and influence on production of beans *Phaseolus vulgaris* L. *Arquivos do Instituto Biologico Sao Paulo* 66, 43-46.

Schaarschmidt, L. (1959) Systematik und Okologie der Tarsonemiden. *Beiträge zur Systematik und Ökologie Mitteleuropäischer Acarina* 1, 5, 713-823.

Sepswasdi, P., Chareonrak, T., Phapoom, V., Pitaksa, S., Chareonrut, S. and Poonyathawon, P. (1991) Chemical control of mungbean insect pests in rice-based cropping systems. In: *Proceedings of the Mungbean Meeting 90*. Bangkok, Thailand, Tropical Agriculture Research Centre, pp. 269-275.

Shoukry, I.F.I., Kandeel, M.M.H. and El-Bashier, Z. (1990) Incidence of acarofauna in the house dust in Sharqiya Governorate, Egypt. *Bulletin de la Societe Entomologique d'Egypte* 69, 183-195.

Titov, D.A. (1986) Mites on strawberry. *Zashchita Rastenii* 1986(9), 48.

Tuovinen, T. (2000) Integrated control of the strawberry mite (*Phytonemus pallidus*) in the Nordic multi-year growing system. *Acta Horticulturae* 525, 389-391.

Tuovinen, T. (2002) Biological control of strawberry mite: a case study. *Acta Horticulturae* 567(2), 671-674.

Tusnadi, C.K. and Kerenyi-Nemestothy, K. (1989) New acaricides against the mite Tarsonemus pallidus sensu stricto Banks (Acari: Tarsonemidae) on Saintpaulia. *Novenyvedelem* 25, 393-399.

Vargas, C. and Ochoa, R. (1990) Medios de cultivo en laboratorio contaminados por *Tarsonemus bilobatus* Suski (Acari: Tarsonemidae) y redescripcion de la especie. *Manejo-Integrado de Plagos* 18, 19-23.

Vieira, M.R. and Chiavegato, L.G. (1999) Biology of *Polyphagotarsonemus latus* (Banks) (Acari: Tarsonemidae) on lemon (*Citrus limon* Burm). *Anais da Sociedade Entomologica de Brasil* 28(1), 27-33.

Wang, D.S., Kuang, K.Y., Wu, S.C., Zhu, Z.Y., Yuan, Y.D., Chen, Y.L. and Yang, X.Q. (2000) The occurrence and control of dominant insect pests and diseases on sweet pepper in advanced greenhouses. *Acta Agriculturae Shanghai* 16(Supplement), 10-16.

Welch, N.C., Pickel, C., Walsh, D. and Nouhuys, S. van (1989) Cyclamen mite control in strawberries. *California Agriculture* 43, 14-15.

Wu, X.-X., Hu, D.-X. and Shen, Z.-R. (2000) Studies on phoresy of the broad mite, *Polyphagotarsonemus latus* (Banks), by the greenhouse whitefly, *Trialeurodes vaporariorum*, under different environmental conditions. *Acta Entomologica Sinica* 43, 157-163.

Xie, M., Cheng, H.K. and Zhao, J.H. (1992) Use of liuyangmycin to control yellow mite, *Polyphagotarsonemus latus* (Acari: Tarsonemidae) infesting green pepper. *Chinese Journal of Biological Control* 8, 29-32.

Eriophyoid Mites

7

7.1 Introduction

Eriophyoid mites belong to the superfamily Eriophyoidea of the order Prostigmata. These worm-like or fusiform mites are too small to be seen by the naked eye, but they cause many forms of plant abnormalities such as galls, leaf blisters and rusts, which are very obvious and indicative of the presence of these mites. Most species are monophagous and many species are limited to plant species within a single genus, with few exceptions. Most species cause little harm to their host plants. However, some rust mites and gall mites are important pests on economic plants.

The Eriophyoidea is a large superfamily of worldwide distribution. Over 3,000 species belonging to over 250 genera are known in the world and ten of thousands of more species are yet to be discovered. The superfamily consists of three families: Eriophyidae, Phytoptidae and Diptilomiopidae. About three-quarters of the described species of the superfamily belong to the Eriophyidae.

References & suggested further reading. Jeppson *et al.* (1975); Amrine and Stasny (1994); Lindquist *et al.* (1996); Oldfield (1996).

7.2 Morphological characters

Eriophyoid mites are tiny annulate mites ranging in size from 90-350 μm. They are worm-like or fusiform and are often pale yellowish to brownish in colour when alive.

The gnathosoma is modified as a median rostrum enclosed by indistinctively segmented lateral palpi. The movable digits of the chelicerae are whip-like and lie in a dorsal groove of the palpal bases; they are very long in the Diptilomiopidae and relatively short in the Eriophyidae and Phytoptidae.

There are no eyes on the prodorsum. Stigmata and peritremes are also absent. The propodosoma often bears a dorsal shield with up to five setae. The dorsal shield may have a median line, a pair of admedian lines and a pair of submedian lines. The subcapicular setae, when present, may each arise from a tubercle. The hysterosoma has a maximum of nine pairs of setae. The genital pore in the female is transverse and located behind coxae II, with an anteriorly hinged flap.

There are only two pairs of legs on the propodosoma. Both pairs of legs terminate in a feather-like empodium and are without true claws.

References & suggested further reading. Jeppson *et al.* (1975); Lindquist (1996).

7.3 Life history and biology

The life cycle passes through the egg, larva, nymph and adult stages. As in spider mites, males develop slightly faster than females. Development is temperature-dependent. Both females and males complete their life cycle in about a week around 25°C. In temperate regions, the life cycle may be more complex due to the presence of an overwintering 'deutogyne' female.

Mating is indirect: males deposit spermatophores on host plants and females then pick them up; the genital flap in the female presses the spermatophore into the body and crushes it, releasing the sperms which fertilize the eggs. Females lay up to three eggs per day for up to a month, with a total of up to 87 eggs per female.

For mites as small as eriophyoids, dispersal is mainly passive via the wind. They may also be spread by insects, birds or the movement of infested plants.

References & suggested further reading. Jeppson *et al.* (1975); Sabelis and Bruin (1996).

7.4 Species important in greenhouses

Most eriophyoid mites attack outdoor plants. Only a couple of them are of significance in greenhouses. *Aculops lycopersici* (Massee) is a pest of greenhouse tomatoes, although its importance is sometimes not as great as *T. urticae* or *P. latus*. *Epitrimerus alinae* Liro is occasionally found on greenhouse grown *Chrysanthemum* in the UK.

7.4.1 *Aculops lycopersici* (Massee)

7.4.1.1 Common Name

This species is most commonly known as the tomato russet mite. It is also known as the tomato rust mite.

7.4.1.2 Distribution And Host Plants

This species is a common pest on tomatoes worldwide. Its occurrence and damage are increasing on greenhouse tomatoes in many European countries: e.g. Israel, Italy, France, Spain, Germany, Moldovia, Bulgaria and Hungary. This species is exceptionally polyphagous for an eriophyid. It is also found on aubergine, tobacco, potato, *Datura* and many other Solanaceae (with the exception of *Convolvulus* species), but only occasionally on chillies and petunia.

References & suggested further reading. Smith (1955); Jeppson *et al.* (1975); Meyer (1981); Kerenyine Nemestothy and Budai (1985); Vacante (1985); Perring and Farrar (1986); Berlinger *et al.* (1988); Trottin-Caudal *et al.* (1989); Izhevskii (1992); Arno *et al.* (1994); Atanasov *et al.* (1995); Manzaroli and Benuzzi (1995); Castagnoli *et al.* (1998); Leuprecht (2000).

7.4.1.3 Appearance And Damage Symptoms

This minute mite is orange-yellow in colour. Because of its small size, it is very difficult to see on plants. Symptoms are therefore useful for recognizing its presence. The infestation of tomato plants starts from the lower leaves, which curl up and turn silvery on the lower surfaces. Later, infested leaves turn brown, withered and paper-like, but they do not wilt as they do when they are diseased by the late blight. As the mite population increases, they spread upwards over the stem and upper leaves. The lower parts of the stem lose their surface hairs before damage appears on the upper leaves. The infested stem changes from green to brown and its surface cracks longitudinally, displaying a characteristic rusty-brown russeted aspect. As infestation continues, the browning of leaves and stems increases and some injured leaves start to drop. With the defoliation of the plant, fruit production is reduced and fruits sunburnt, although they do not usually russet. In very heavily infested plants, young leaves may be distorted and fruits bronzed when attacked.

Mite feeding reduces the rate of photosynthesis by host plants. This is caused by the destruction of the guard cells and subsequent reduction of leaf gas exchange.

References & suggested further reading. Jeppson *et al.* (1975); Meyer (1981); Royalty and Perring (1989).

7.4.1.4 Diagnostic Characters

The body is fusiform and robust. Females are 150-180 μm in length. The prodorsal shield has a broad and short anterior lobe, which is topped by a transverse line that extends along the side of the prodorsal shield to the posterior margins. These lines have lateral branches, forming characteristic cell-like ornamentation (Fig. 7.1B). The median line is faintly seen only on the posterior two-thirds of the shield, with a pair of longer admedian lines running almost in parallel. The anterior and posterior portions of the admedian line curves outward. The dorsal idiosomal setae are short. The hysterosoma has 27 tergites with indistinct, elongate microtubercles and 60 sternites with bead-like microtubercles (Fig. 7.1A and E). The female genitalia have faint basal granules and about ten longitudinal ribs (Fig. 7.1C). Each leg terminates in a four-rayed feather-like empodium (Fig. 7.1A, D).

References & suggested further reading. Jeppson *et al.* (1975); Meyer (1981); Keifer *et al.* (1982).

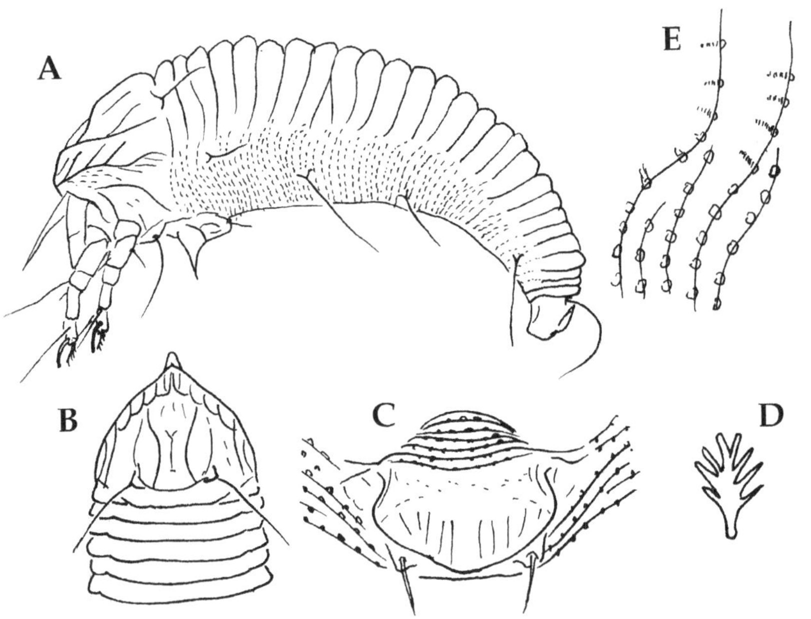

Fig. 7.1 *Aculops lycopersici* (Massee). A, habitus of adult female, lateral view; B, prodorsum and anterior tergites; C, female genitalia; D, feather-like empodium on leg tarsus; E, details of cuticle structure, showing indistinct, elongate microtubercles on the tergites and bead-like microtubercles on the sternites (after Keifer *et al.*, 1982).

7.4.1.5 Life History and Biology

The life cycle of this mite passes through egg, larva, nymph and adult stages. The generation time is about one week at 21-25°C (Table 7.1). Males develop slightly faster than females. Females live for several weeks and lay ten to 53 eggs. Fertilized eggs produce both males and females, whereas unfertilized eggs give rise to males only.

The optimal conditions for mite development are 21-27°C and 30% RH, conditions that greenhouses often provide. At higher temperatures, lower humidity is required; high humidity can cause increased mite mortality.

References & suggested further reading. Bailey and Keifer (1943); Rice and Strong (1962); Abou-Awad (1979); Baradaran-Anaraki and Daneshvar (1992).

Table 7.1 Life history features of *Aculops lycopersici* (Massee).

	Bailey and Keifer (1943) at 21°C	Rice and Strong (1962) at 21°C	Abou-Awad (1979) at 25°C
Egg	2.0 days	-	2.3 (2.3)* days
Larva	1.0 day	-	1.0 (0.8) days
Nymph	2.0 days	-	1.8 (1.5) days
Preoviposition	2.0 days	-	2.2 days
Total life cycle	7.0 days	6.5 days	7.3 days
Longevity		47.5 days	22.1 (16.5) days
Fecundity	15 eggs	10-53 eggs	16 eggs

*Value for female with that for males in parentheses.

7.4.1.6 Control and Management

Biological control

Natural enemies of the tomato russet mites are many and include both predatory insects and mites.

Two species of predatory thrips, *Leptothrips mali* (Fitch) and *Scolothrips sexmaculatus* (Pergande), are known to be associated with this mite and the former actively feeds on *A. lycopersici* but is hindered by glandular hairs on tomato. Their predation effects on *A. lycopersici* are unknown.

A tydeid mite, *Pronematus ubiquitus* (McGregor), is known to be associated with *A. lycopersici* but nothing is known about its effectiveness as a predator of *A. lycopersici*. Another tydeid mite, *Homeopronematus anconai* (Baker), however, is known to feed on all stages of the tomato russet mite but it can not develop to the adult stage when fed solely on this prey; it needs to feed on alternative foods such as fungal spores to complete development. The ability to feed on alternative foods may be an advantage because its densities may be sustained by alternative foods when eriophyoid mites are low in density.

A stigmaeid mite, *Agistemus exsertus* Gonzales, has been evaluated in the laboratory with promising results, and is recommended for releases on to mite-infested plants for biological control. However, this species is not commercially available.

Several species of phytoseiids have been evaluated as predators of the tomato russet mite. *Phytoseiulus persimilis* does not feed on *A. lycopersici*. *Neoseiulus cucumeris* develops successfully on *A. lycopersici*, but can not reproduce on it. *Euseius victoriensis* consumes a couple of mites per day and can reproduce on it, but at very low rates (0.1 eggs per day). *Euseius concordis* (Chant) develops and reproduces well on the tomato russet mite and is naturally associated with the mite on tomato plants. However, it is hindered by the webbing produced by spider mites which occur with the eriophyid on tomato plants. Only *Neoseiulus fallacis* seems to be a good biocontrol agent of this mite. It attacks all stages of the pest, survives well (92%), develops rapidly (6.3 days at 22°C), and reproduces well on *A. lycopersici*. This commercially available predator should be effective when released at a rate of 1,000 per acre when the tomato seedlings are transplanted into the greenhouse.

Pathogenic bacteria such as BT (*Bacillus thuringiensis*) may be a new biological weapon for mite control. For example, an exotoxin from *Bacillus thuringiensis thuringiensis* shows promising experimental results in the biocontrol of *A. lycopersici* on tomato plants in Uzbekistan.

Chemical control

Over 70 chemicals were tested for control of this mite during 1940-1985. Those tested since 1980 that provide good control include: binapacryl, bromopropylate, cyhexatin, fenvalerate, flubenzimine, fluvalinate, ethion, dichlorvos, dicofol, formothion, karathane, lambda-cyhalothrin, methamidophos, milbex, profenofos, propargite pyridaphenthion and zineb. Abamectin and sulphur are very effective and widely used, and may facilitate biological control of other pests on tomato, especially *T. cinnabarinus* by predatory mites.

Various kinds of oils are also more environmentally friendly than synthetic pesticides and can be used effectively against the tomato russet mite.

Integrated control

In integrated pest control, it is essential to monitor the pests. It is useful to regularly search for bronzing on lower leaves and stems and then check damaged leaves and other leaves immediately above them for mites. Damage is typically first observed when green fruits reach 5 cm; rarely is it first observed when more than 25% of the fruit are ripe. For effective control, treatment is necessary immediately when damage symptoms begin to spread. A thorough clean-up of the greenhouse and good sanitation will help to prevent the carry-over of this mite.

Some varieties of tomato are known to be less susceptible to tomato russet mite damage and may be used in IPM programmes.

References & suggested further reading. Anderson (1954); Abou-Awad (1979); Moraes and Lima (1983); Hessein and Perring (1986, 1988); Haji *et al.* (1988); Osman and Zaki (1986); Perring and Farrar (1986); Royalty and Perring (1987); Berlinger *et al.* (1988); Undurraga and Dybasc (1988); James (1989); Cheremushkina *et al.* (1991); Costilla (1991); Baradaran-Anaraki and Daneshvar (1992); Kamau *et al.* (1992); Monkman (1992); Brough *et al.* (1994); Manzaroli and Benuzzi (1995); Brodeur *et al.* (1997); Calpas (1998); Singh *et al.* (2000); Sulaimanov *et al.* (2000).

7.4.2 *Epitrimerus alinae* Liro

7.4.2.1 Common Name

This species is most commonly known as the chrysanthemum leaf rust mite.

7.4.2.2 Distribution And Host Plants

This species was originally described by Liro from chrysanthemum plants in Finland. It has been known to attack *Chrysanthemum* in greenhouses in the UK and can cause considerable damage to the plant.

References & suggested further reading. Liro (1941); Vernon (1957); Alford (1994).

7.4.2.3 Appearance And Damage Symptoms

Because of its small size, this species is very difficult to see on plants and its presence is revealed by the symptoms. This species feeds among the hairs, mostly 7.6-15.2 cm below the florets, on the chrysanthemum stems. Mite feeding causes the stem of the plant to become russeted. This often occurs near upper petioles. Infested leaves may wilt and fall prematurely. When the stems are severely damaged, mites tend to move to the sepals and florets. Mites are also found on the upper leaf surface, but in much lower numbers.

References & suggested further reading. Vernon (1957); Hussey *et al.* (1969); Alford (1994).

7.4.2.4 Diagnostic Characters

The adult female is fusiform, 160 µm long. The prodorsal shield is more or less triangular in shape, 38 µm long, without cell-like ornamentation in *A. lycopersici* (Fig. 7.2B). The idiosoma has 43-45 tergites (Fig. 7.2A; cf: 27 tergites in *A. lycopersici* in Fig. 7.1.A). The feather-like empodium on each leg is four-rayed.

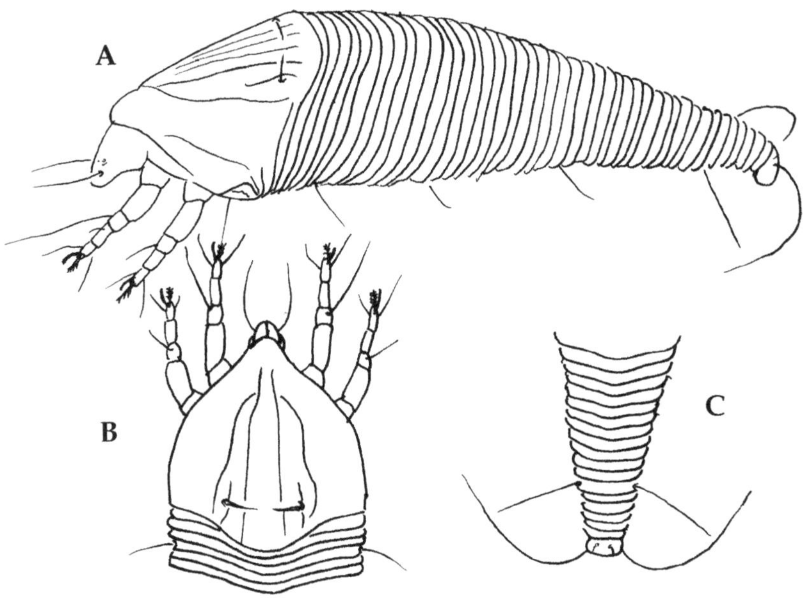

Fig. 7.2 *Epitremerus alinae*. A, habitus of adult female, lateral view; B, prosoma, dorsal view; C, posterior opisthosoma, dorsal view (after Liro, 1941).

7.4.2.5 Life History And Biology

Little is known about the life history and biology of this species. It is probably similar to other species in having a short life cycle in greenhouses. On chrysanthemum stems, its density can reach up to 130 live mites per top 15.2 cm.

References & suggested further reading. Vernon (1957); Lindquist (1996).

7.4.2.6 CONTROL AND MANAGEMENT

There are no reports on the association of predatory mites with *E. alinae*. However, generalist predators effective against other eriophyids may be tried on this species.

Chlorobenzilate and sulphur are not effective, but metasystox is effective against it.

Removal of infested stems and foliage may help to prevent its spread and growth. Other common sense methods may be tried as well (Table 1.5).

References & suggested further reading. Vernon (1957); Lindquist (1996).

7.4.3 Other eriophyoid mites

7.4.3.1 *Aceria lycopersici* (Wolffenstein)

This species is commonly known as the tomato erineum mite. It is distributed widely in the tropical areas and may occur in greenhouses in the temperate regions. It feeds on the tomato, aubergine and other solanaceous plants. It is generally an outdoor pest and has not been reported as of significance on tomato plants in greenhouses.

The feeding of this mite induces hairlike growth (erineum) on stems and leaves, which gives the plant a silvery-white appearance. This will allow easy separation of this species from the tomato russet mite. Morphologically, this species is distinguished from the tomato russet mite by the presence of an almost complete median line on the prodorsal shield (cf: it is faintly seen only on the posterior two-thirds of the shield in the tomato russet mite).

Little is known about the biology and control of this species. It is known to thrive in dry seasons in the tropics. The phytoseid predator, *Euseius gossipi* (El-Badry) can feed on this mite and complete development, although developmental rates are slower than on a diet of spider mites or pollen. Sulfur is effective in controlling this species.

References & suggested further reading. Massee (1939); Jeppson *et al.* (1975); Abou-Awad (1983).

7.4.3.2 *Paraphytoptus chrysanthemi* Keifer

This species is known as the chrysanthemum rust mite in the USA and chrysanthemum semi-bud mite in the UK. It attacks various species of *Chrysanthemum*, but is not known to attack other plants, nor is it distributed in other countries. It is known to cause serious injury to outdoor chrysanthemums and may occur in greenhouses.

Mites feed among hairs on the under surface of leaves and green stems. Mite feeding causes the shortening of the stems, stunting and curling of apical leaves and clustering of stems in some brooming. Some attacked plants have reddish leaves and no flowers or few deformed flowers. Infested flowers become vegetative and appear leaf-like.

Adults are fusiform and light-yellowish in colour. This species is distinguished from *E. alinae* by its symptom. Morphologically, this species is easily separated from *E. alinae* by its anterior three quarters of the idiosoma without differentiation between tergites and sternites.

Nothing is known about the biology and control of this species. For possible biological control, generalist mite predators effective against other eriophyids may be tried on this species.

References & suggested further reading. Breakey and Batchelor (1950); Miles (1964); Jeppson *et al.* (1975).

7.4.3.3. *Aceria* on carnations

Three species of *Aceria* are known from carnations and may occur in greenhouse grown carnations.

Aceria dianthi (Lindroth) is only known in Finland from the carnation *Dianthus deltoides*. It is a leaf vagrant and causes a stunting of the plants.

Aceria paradiathi Keifer is known from various species of *Dianthus* in Europe, Argentina and the USA. The mites prefer lower parts of the plant and feed between leaf bases and stems, producing a greasy and distorted appearance and stunted and discoloured plant.

Aceria georghioui (Keifer) is known from carnations in Cyprus and California (USA). Infested plants become discoloured and distorted.

The three species are tiny and are often recognized by the symptoms on host plants. Morphologically, they are also easy to distinguish. The feather-like empodium is five-rayed in *A. dianthi*, six-rayed in *A. paradianthi* and seven-rayed in *A. georghioui.*

Little is known about the economic significance, biology and control of these mites. *A. georghioui* is known to be susceptible to diazinon, chlorobenzilate, dicofol and parathion, but repeated applications are needed to prevent mite injury. Generalist predatory mites such as *Neoseiulus cucumeris* may be tried against these mites.

References & suggested further reading. Hussey *et al.* (1969); Jeppson *et al.* (1975); Meyer (1996).

7.5 References cited

Abou-Awad, B.A. (1979) The tomato russet mite, *Aculops lycopersici* (Massee) (Acari: Eriophyidae) in Egypt. *Anzeiger für Schädlingskunde, Pflanzenschutz, Umweltschutz* 52, 153-156.

Abou-Awad, B.A. (1983) *Amblyseius gossypi* (Acarina: Phytoseiidae) as a predator of the tomato erineum mite, *Eriophyes lycopersici* (Acarina: Eriophyidae). *Entomophaga* 28, 263-266.

Alford, D.V. (1994) *A Colour Atlas of Pests of Ornamental Trees, Shrubs and Flowers.* Wolfe Publishing Ltd., London. 448 pp.

Amrine, J.W. Jr and Stasny, T.A. (1994) *Catalog of the Eriophyoidea (Acarina: Prostigmata) of the World.* Indira Publishing House, West Bloomfield, Michigan, USA. 798 pp.

Anderson, L.D. (1954) The tomato russet mite in the United States. *Journal of Economic Entomology* 47, 1001-1005.

Arno, J., Moliner, J. and Gabarra, R. (1994) Integrated pest control of early greenhouse tomato in the Isle of Menorca. *Boletin de Sanidad Vegetal Plagas* 20(2), 501-509.

Atanasov, N.D., Fernandez-Munoz, R., Cuartero, J. and Gomez-Guillamon, M.L. (1995) Integrated control of mites on tomatoes. First international symposium on solanacea for fresh market, Malaga, Spain, 28-31 March, 1995. *Acta Horticulturae* 412, 546-550.

Bailey, J.S. and Keifer, H.H. (1943) The tomato russet mite, *Phyllocoptes destructor* Keifer: its present status. *Journal of Economic Entomology* 36, 702-712.

Baradaran-Anaraki, P. and Daneshvar, H. (1992) Studies on the biology and chemical control of tomato russet mite, *Aculops lycopersici* (Acari: Eriophyidae), in Varamin. *Applied Entomology and Phytopathology* 59(1-2), 25-27.

Berlinger, M.J., Dahan, R. and Mordechi, S. (1988) Integrated pest management of organically grown greenhouse tomatoes in Israel. *Applied Agricultural Research* 3(5), 233-238.

Breakey, E.P. and Batchelor, G.S. (1950) Phyllody of chrysanthemum and the eriophyid mite, *Paraphytoptus chrysanthemi* Keifer. *Annals of the Entomological Society of America* 43, 492-494.

Brodeur, J., Bouchard, A. and Turcotte, G. (1997) Potential of four species of predatory mites as biological control agents of the tomato russet mite, *Aculops lycopersici* (Massee) (Eriophyidae). *Canadian Entomologist* 129, 1-6.

Brough, E.J., Elder, R.J. and Beavis, C.H.S. (1994) *Managing Insects and Mites in Horticultural Crops.* Department of Primary Industries Queensland. 231 pp.

Calpas, J. (1998) Tomato Russet Mite. *Greenhouse Coverings*, October 1998. http://www.agric.gov.ab.ca/crops/hort/grnhouse/9810-russet_mite.html

Castagnoli, M., Liguori, M., Nannelli, R. and Simoni, S. (1998) Preliminary survey on the mite fauna of tomato in Italy. *Redia* 81, 45-54.

Cheremushkina, N.P., Aramov, N.Kh., Makarenkova, A.A. and Golyshin, N.M. (1991) The rust mite of tomato. *Zashchita Rastenii* 1991(11), 44-45.

Costilla, M.A. (1991) The mite *Aculops lycopersici* (Massee, 1937) (Acari: Eriophyidae) responsible for tomato russet in the North East of Argentina. *Revista Industrial y Agricola de Tucuman* 68, 83-90.

Haji, F.N.P., Moraes, G.J. de, Lacerda, C.A. de and Neto, R.S. (1988) Chemical control of the tomato russet mite *Aculops lycopersici* (Massee, 1937). *Anais da Sociedade Entomologica do Brasil* 17, 437-442.

Hessein, N.A. and Perring, T.M. (1986) Feeding habits of the Tydeidae with evidence of *Homeopronematus anconai* (Acari: Tydeidae) predation on *Aculops lycopersici* (Acari: Eriophyidae). *International Journal of Acarology* 12, 215-221.

Hessein, N.A. and Perring, T.M. (1988) The importance of alternative foods for the mite *Homeopronematus anconai* (Acari: Tydeidae). *Annals of the Entomological Society of America* 81, 488-492.

Hussey, N.W., Read, W.H. and Hesling, J.J. (1969) *The Pests of Protected Cultivation: The Biology and Control of Glasshouse and Mushroom Pests.* Edward Arnold (Publishers) Ltd., London. 404 pp.

Izhevskii, S.S. (1992) New pests of greenhouse plants. *Zashchita Rastenii Moskva*, 1992(12), 26-27.

James, D.G. (1989) Influence of diet on development, survival and oviposition in an Australian phytoseiid, *Amblyseius victoriensis* (Acari: Phytoseiidae). *Experimental & Applied Acarology* 6, 1-10.

Jeppson, L.R., Baker, E.W. and Keifer, H.H. (1975) *Mites Injurious to Economic Plants.* University of California Press, Berkeley, California, 614 pp.

Kamau, A.W., Mueke, J.M. and Khaemba, B.M. (1992) Resistance of tomato varieties to the tomato russet mite, *Aculops lycopersici* (Massee) (Acarina: Eriophyidae). *Insect Science and its Application* 13, 351-356.

Keifer, H., Baker, E.W., Kono, T., Delfinado, M. and Styer, W. (1982) An illustrated guide to plant abnormalities caused by eriophyid mites in North America. *United States Department of Agriculture Agricultural Research Service Agriculture Handbook* 573, 1-178.

Kerenyine Nemestothy, K. and Budai, C. (1985) Leaf mites (Acari, Eriophyidae) as new greenhouse pests. *Novenyvedelem* 21, 220.

Leuprecht, B (2000) Occurrence of tomato rust mite in tomatoes under glass. *Gemuse Munchen* 36(3), 26-27.

Lindquist, E.E. (1996) External anatomy and notation of structures. In: Lindquist, E.E., Sabelis, M.W. and Bruin, J. (eds) *Eriophyoid Mites: Their Biology, Natural Enemies and Control. World Crop Pests,* Volume 6. Elsevier, Amsterdam. pp. 3-31.

Lindquist, E.E., Sabelis, M.W. and Bruin, J. (eds) (1996) *Eriophyoid Mites: Their Biology, Natural Enemies and Control. World Crop Pests,* Volume 6. Elsevier, Amsterdam. 790 pp.

Liro, J.I. (1941) Uber neue und seltene Eriophyiden (Acarina). *Annales zoologici Societatis zoologico-botanicae fennicae* Vanamo 8, 1-68.

Manzaroli, G. and Benuzzi, M (1995) Protected tomato: biological and integrated control. *Colture Protette* 24(1), 41-47.

Massee, A.M. (1939) A species of gall mite (Eriophyidae) injurious to tomato. *Annals and Magazines of Natural History Series* 2, 3, 617-619.

Meyer, M.K.P. (Smith) (1981) Mite pests of crops in Southern Africa. *Science Bulletin Department of Agriculture and Fisheries Republic of South Africa* 397, 1-92.

Meyer, M.K.P. (1996) Ornamental flowering plants. In: Lindquist, E.E., Sabelis, M.W. and Bruin, J. (eds) *Eriophyoid Mites: Their Biology, Natural Enemies and Control. World Crop Pests*, Volume 6. Elsevier, Amsterdam, pp. 641-650.

Miles, P.M. (1964) Chrysanthemum semi-bud mite in Wales. *Entomologist Monthly Magazine* 100, 97.

Monkman, K.D. (1992) Tomato russet mite - Acari: Eriophyoidea. *Monthly Bulletin Department of Agricuture, Fisheries and Parks, Bermuda* 63(3), 19-21.

Moraes, G.J. de and Lima, H.C. (1983) Biology of *Euseius concordis* (Chant), a predator of the tomato russet mite. *Acarologia* 24, 251-255.

Oldfield, G.N. (1996) Diversity and host plant specificity. In: Lindquist, E.E., Sabelis, M.W. and Bruin, J. (eds) *Eriophyoid Mites: Their Biology, Natural Enemies and Control. World Crop Pests*, Volume 6. Elsevier, Amsterdam, pp. 199-216.

Osman, A.A. and Zaki, A.M. (1986) Studies on the predation efficiency of *Agitemus exsertus* Gonzalez (Acarinaa, Stigmaeidae) on the eriophyid mite, *Aculops lycopersici* (Massee). *Anzeiger für Schädlingskunde, Pflanzenschutz, Umweltschutz* 59, 135-136.

Perring, T.M. and Farrar, C.A. (1986) Historical perspective and current world status of the tomato russet mite (Acari: Eriophyidae). *Miscellaneous Publications of the Entomological Society of America* 63, 1-19.

Rice, R.E. and Strong, F.E. (1962) Bionomics of the tomato russet mite, *Vasates lycopersici* (Massee). *Journal of Enonomic Entomology* 55, 431-435.

Royalty, R.N. and Perring, T.M. (1987) Comparative toxicity of acaricides to *Aculops lycopersici* and *Homeopronematus anconai* (Acari: Eriophyidae, Tydeidae). *Journal of Economic Entomology* 80, 345-351.

Royalty, R.N. and Perring, T.M. (1989) Reduction in photosynthesis of tomato leaflets caused by tomato russet mite (Acari: Eriophyidae). *Environmental Entomology* 18, 256-260.

Sabelis, M.W. and Bruin, J. (1996) Evolutionary ecology: life history patterns, food plant choice and dispersal. In: Lindquist, E.E., Sabelis, M.W. and Bruin, J. (eds) *Eriophyoid Mites: Their Biology, Natural Enemies and Control. World Crop Pests*, Volume 6. Elsevier, Amsterdam, pp. 329-366.

Singh, P., Beattie, G.A.C., Clift, A.D., Watson, D.M., Furness, G.O., Tesoriero, L., Rajakulendran, V., Parkes, R.A. and Scanes, M. (2000) Petroleum spray oils and tomato integrated pest and disease management in southern Australia. *General and Applied Entomology* 29, 69-93.

Smith, W.A. (1955) The tomato mite. *Queensland Agricultural Journal* 81, 90-91.

Sulaimanov, B., Rashidov, M.I. and Kimsanbaev, Kh. (2000) Exotoxin against the rust mite on solanaceous crops. *Zashchita i Karantin Rastenii* 2000(3), 46.

Trottin-Caudal, Y., Grasselly, D., Trapateau, M. and Villevieille, M. (1989) The principal mites found on Solanaceae and Cucurbitaceae grown under glass in

France. *Infos Paris* 1989(57), 9-13

Undurraga, J.M. and Dybas, R.A. (1988) Abamectin, novel naturally-derived insecticide/acaricide for integral key pest controls on pears and tomatoes. In: *Brighton Crop Protection Conference, Pests and Diseases*. British Crop Protection Council, Thornton Heath, UK, pp. 205-210.

Vacante, V. (1985) Acari present in horticulture and floriculture in greenhouses of the Ragusa area and considerations on the possibility of chemical, biological and integrated control. *Tecnica Agricola* 37, 299-321.

Vernon, J.D.R. (1957) An eriophyid causing damage to chrysanthemums. *Plant Pathology* 6(3), 111.

Acarid Mites

8

8.1 Introduction

Acarid mites belong to the family Acaridae of the order Astigmata. Most species are fungivorous and commonly occur in stored food products and decaying organic matter. Some species are also facultatively phytophagous. Many other species are associates of various insects, or inhabitants of vertebrate nests.

The Acaridae is a large family of worldwide distribution. About 400 species of acarid mites belonging to some 90 genera are known in the world and many others are yet to be discovered, especially in the tropical areas. The genus *Rhizoglyphus* contains several species of pests attacking root crops. *Acarus* and *Tyrophagus* are the most abundant and economically important mites inhabiting stored food and products. Some *Tyrophagus* species can cause economic damage to plants, including both ornamental flowers and vegetables grown in greenhouses.

References & suggested further reading. Hughes (1976); O'Connor (1982); Diaz *et al.* (2000).

8.2 Morphological characters

Acarids are medium-sized, soft-bodied mites. They are often pale whitish to brownish in colour.

The propodosoma often has a shield-like prodorsal sclerite. The idiosoma often has a sejugal furrow in all stages. There is often a full complement of dorsal idiosomal setae and reductions occur only in some derived lineages. Dorsal setae are often nude or slightly barbed. The coxal apodemes are strongly developed as in the Tarsonemidae (Chapter 6).

The genital opening is located between or behind coxae IV. Most males have para-anal suckers.

The legs are often short, but can be long in some species. They terminate in a well-developed empodial claw, which is connected with a pair of stout parallel sclerites (condylophores). Genu I has two solenidia.

Males and females are often similar in structure with minimal sexual dimorphism, but in some species, males may be homeomorphic and heteromorphic with females. The latter usually have thick body setae and a strongly developed third pair of legs.

References & suggested further reading. Hughes (1976); O'Connor (1982).

8.3 Life history and biology

The life cycle consists of the egg, larva, protonymph, deutonymph, tritonymph and adult stages. The deutonymph may or may not be present, depending on environmental and biotic conditions. The deutonymph is specially known as the hypopus. It is a non-feeding stage adapted for dispersal and resisting adverse environmental conditions. It is often yellow or brown in colour and is very well-sclerotized. It is oval in shape and often dorsoventrally flat or dorsally convex/ventrally concave so that it is easy to attach to its host for dispersal. The ventral opisthosoma is armed with sucker plates for adhering to hosts.

Development from the egg to adult normally takes one to three weeks, depending on temperature. Many acarid mites are very highly fecund. For example, *Rhizoglyphus* females can produce some 500 eggs.

References & suggested further reading. Hughes (1976).

8.4 Species important in greenhouses

Rhizoglyphus robini (Claparède) and *R. echinopus* (Fumouze and Robin) are important pests of bulbs and corms of many ornamental plants. *Tyrophagus putrescentiae* (Schrank), *T. longior* (Gervais), *T. newswanderi* Johnston and Bruce, *T. perniciosus* Zachvatkin and *T. similis* Volgin occasionally attack greenhouse plants. *Mycetoglyphus fungivorus* (Oudemans) is also known to attack greenhouse grown spinach. These species can be separated using keys in Figs 8.1-8.4. *Acarus* is also included in the key because mites of this genus are used to culture predatory mites such as *Neoseiulus cucumeris* and may be released along with predators into greenhouses.

References & suggested further reading. Hussey *et al.* (1969); Czaikowska *et al.* (1988); Nakao (1989); Voigt (1990).

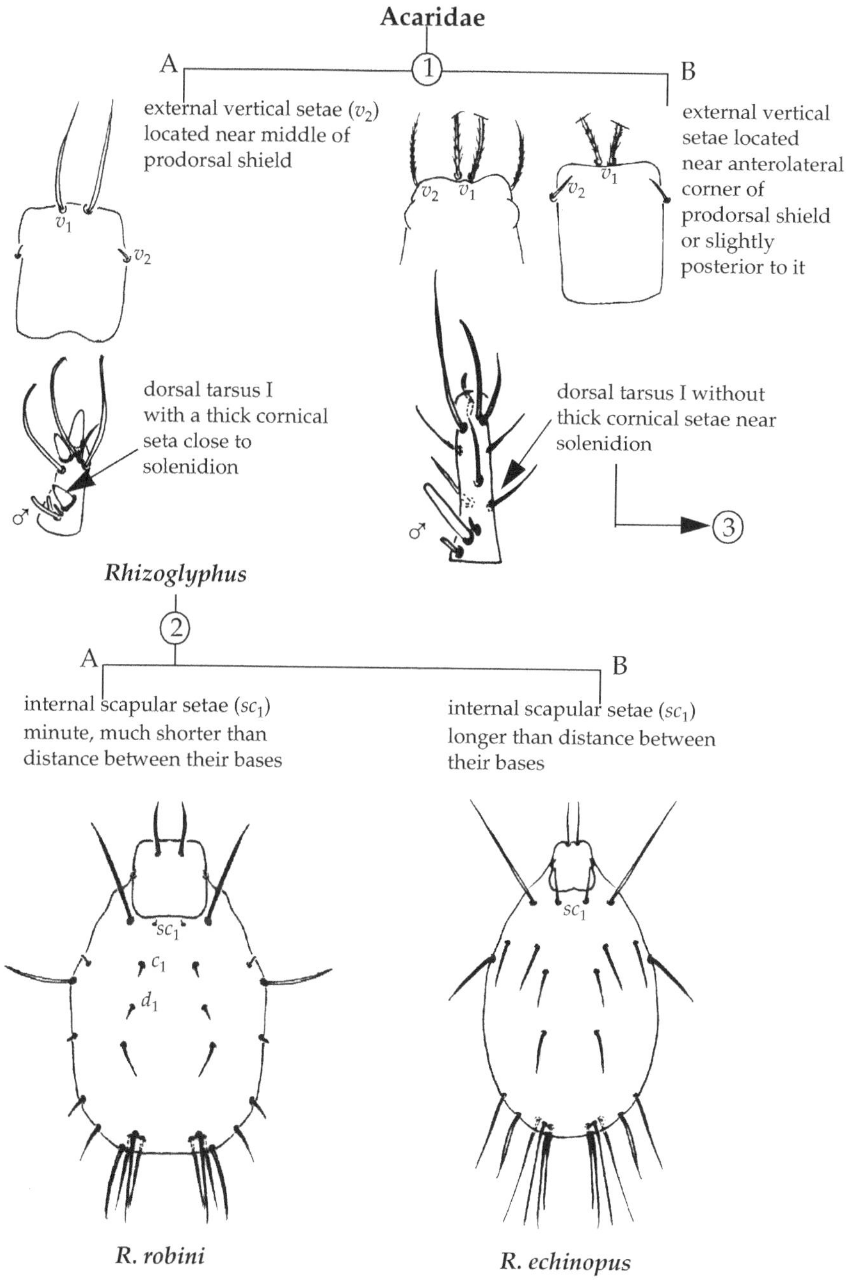

Fig. 8.1 Key to genera and species of greenhouse Acaridae. Part I.

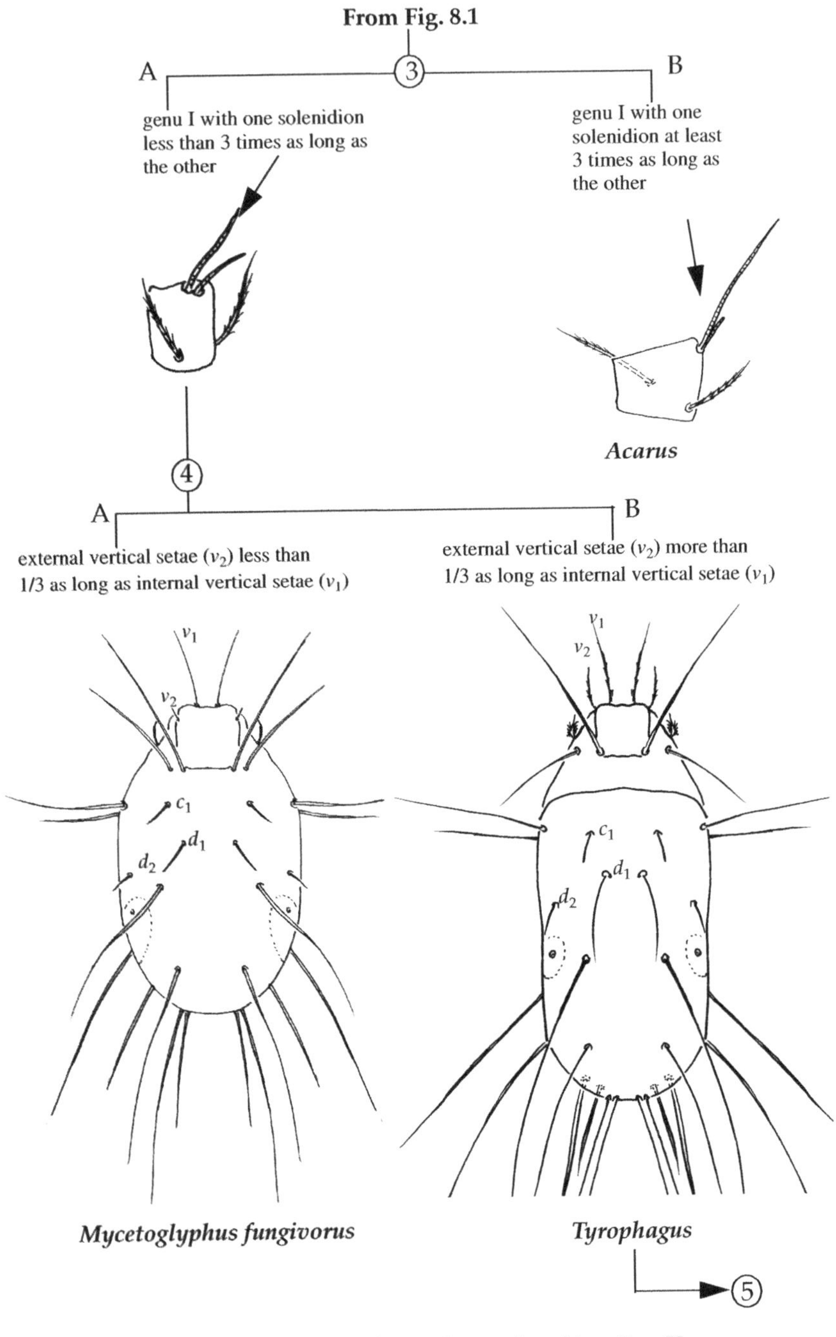

Fig. 8.2 Key to genera and species of greenhouse Acaridae. Part II.

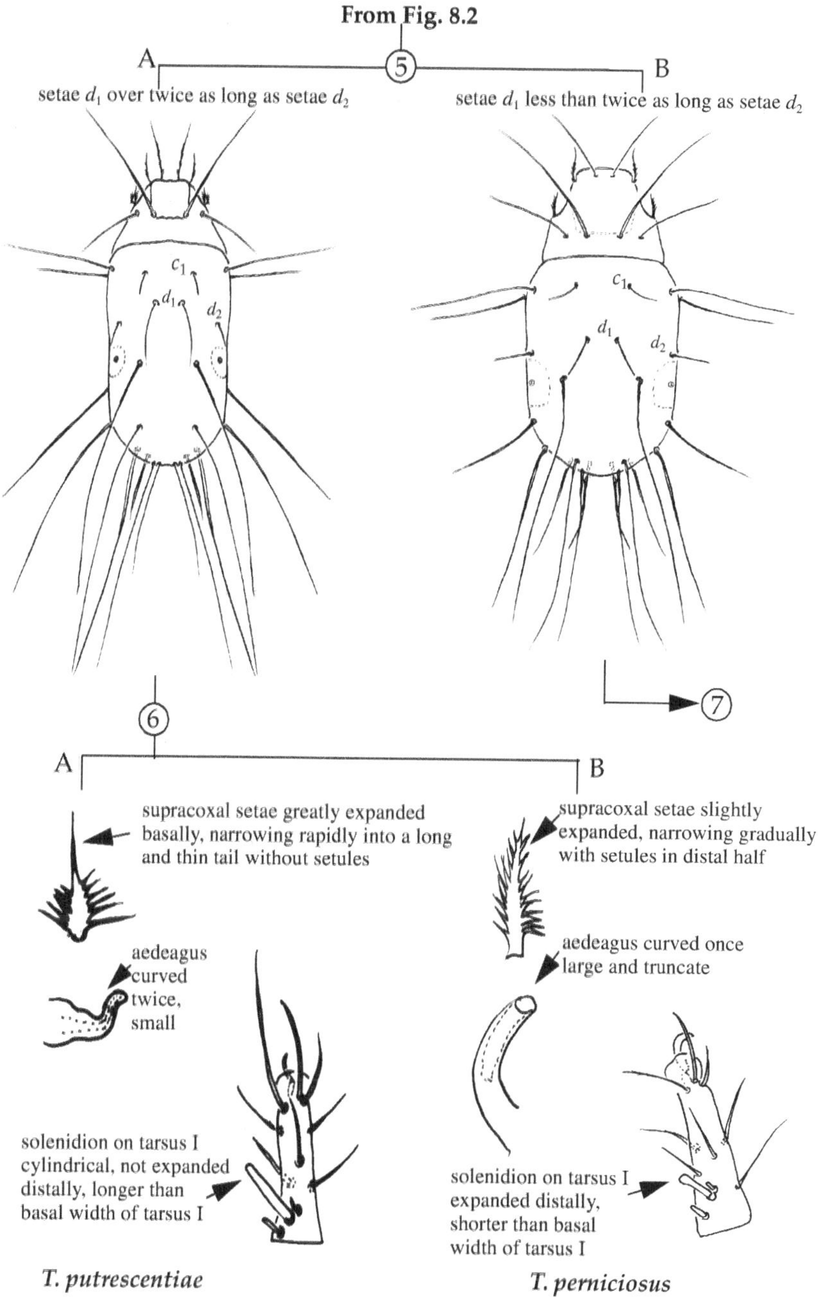

Fig. 8.3 Key to genera and species of greenhouse Acaridae. Part III.

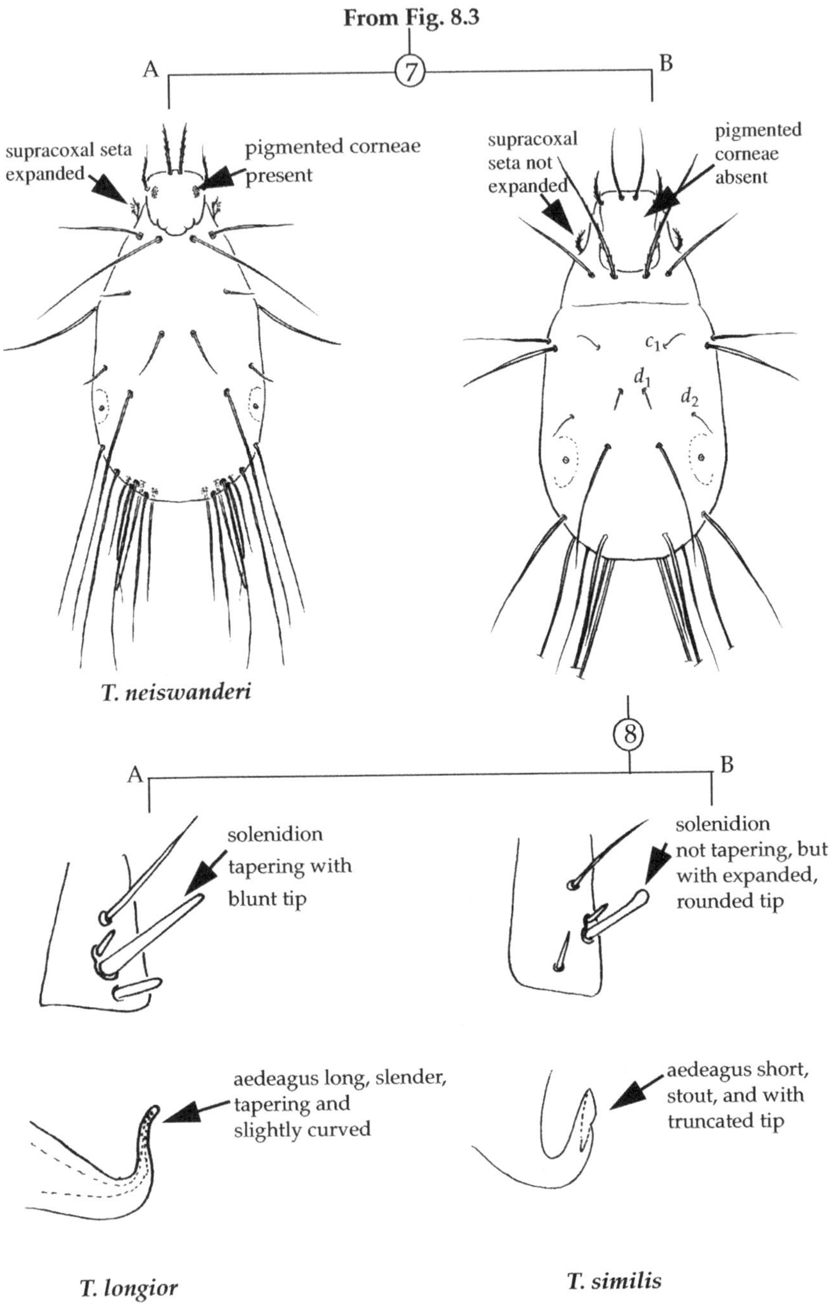

Fig. 8.4 Key to genera and species of greenhouse Acaridae. Part IV.

8.4.1 ***Rhizoglyphus robini*** (Claparède)

8.4.1.1 Common Name

No widely accepted common names have been proposed for this species, although some people call it the robine bulb mite.

8.4.1.2 Distribution And Host Plants

This is a cosmopolitan species. It is common in decaying organic matter and soil. It is often associated with bulbs, roots and seeds of many plant species and is known to attack bulbs of *Narcissus*, *Eucharis*, lilies, orchids, gladioli, hyacinth and tulip, tubers of dahlia and corms of *Freesia* in storage, in greenhouses and in the field.

References & suggested further reading. Hussey *et al.* (1969); Manson (1972); Hughes (1976).

8.4.1.3 Appearance And Damage Symptoms

The body is oval and transculent, with a smooth surface and shiny appearance. The appendages are short and reddish brown. The eggs are oval and slightly over half as long as adult females. They are translucent white. Immature stages are also translucent except the hypopus, which is brownish.

Mite infestation of bulbs and corms starts with penetration through the basal plate or outer skin layers. Bruised bulbs and fungus-infected bulbs allow more rapid mite establishment than healthy ones. On *Fusarium*-infected rakkyo bulbs, *R. robini* achieves exponential population growth after 14 days, whereas on healthy bulbs it takes 60-90 days to achieve exponential population growth. Infested plants have dark brown streaks on roots and develop distorted growing tips and leaves. On an injured bulb, *R. robini* can increase rapidly in number and quickly reduce the whole bulb to a rotten pulp.

References & suggested further reading. Hussey *et al.* (1969); Meyer (1981); Ascerno *et al.* (1983); Okabe and Amano (1991).

8.4.1.4 Diagnostic Characters

The adult female is 600-940 μm long. Dorsal idiosomal setae are short; setae sc_1 are minute (7-25 μm); the first two pairs of dorsomedian setae (c_1, d_1; Fig. 8.1.2A) are shorter than one-third of the distance between their bases. The supracoxal seta is slender, 14-39 μm long. The Grandjean's organ does not have a distinct forked tip. The bursa copulatrix has a relatively small opening at some distance from the anal slit and opens internally into the receptaculum seminis, with two V-shaped projections grouped close together. In the male, the aedeagus is narrower and more cone-shaped than that in *R. echinopus*.

References & suggested further reading. Manson (1972); Hughes (1976); Fan and Zhang (2003).

8.4.1.5 Life History and Biology

The larva is the first mobile stage. A protonymph proceeds the hypopal stage. Before reaching adulthood, there is a tritonymphal stage. Hypopi occur only in a portion of the population, especially under severe conditions. These are flattened individuals without functional mouthparts; they help the mite to disperse and resist adverse conditions. They disperse by attaching to insects such as narcissus flies and sciarids.

Life history parameters are temperature-dependent (Fig. 8.5). They also differ when fed on different food/hosts. The life cycle takes seven to 27 days at 13-26°C excluding the hypopal stage and is nine days at 25°C. The threshold of development is 11.8°C, and the thermal constant is 184.8 day-degrees. Females kept at 27°C lay an average of 400 eggs when reared on garlic. Males live for almost twice as long as females. The intrinsic rate of increase is 0.128 on garlic at 27°C.

References & suggested further reading. Hussey *et al.* (1969); Hughes (1976); Gerson *et al.* (1983); Wang and Lin (1986); Raut and Sarkar (1991); Liu and Tzeng (1994); Ostovan and Kamali (1996).

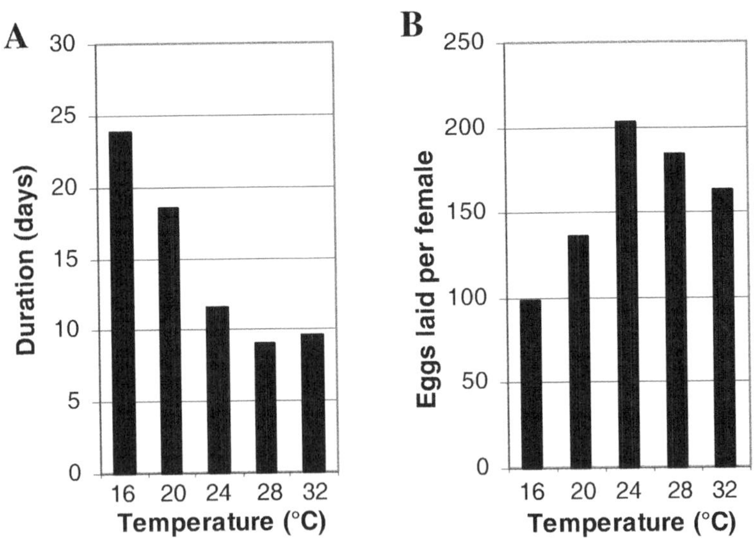

Fig. 8.5 Egg to adult developmental time (A) and total number of eggs laid per female (B) in relation to temperature (drawn from data in Liu and Tzeng, 1994).

8.4.1.6 CONTROL AND MANAGEMENT

Several species of predatory mites have been tested for biological control of *R. robini*. *Hypoaspis aculeifer*, which is commercially available, is effective against this mite in small scale lily bulb propagation at the predator:prey ratio of 1:20, but in large greenhouse plots at the ratios of 1:2 to 1:5. However, as many as three predators per bulb mite are needed for the elimination of the bulb mite population. In small scale lily bulb propagation, the ascid mite *Lasioseius bispinosus* is also able to control the bulb mite, so is the parasitid mite *Parasitus fimetorum*, but for the latter only when the growing medium is peat. Another ascid mite *Protogamasellus minutus* and a digamasellid mite *Dendrolaelaps sayedi* also feed on *R. robini*, but their effectiveness against this mite is unknown.

This species can be controlled using pirimiphos-methyl combined with cultural measures such as hot water treatment (2 h at 39° or 41°C) and storage of the bulbs at –2°C. In peat growing media, a combination of hot water treatment and the release of *H. aculeifer* is also effective against the bulb mite in lilies during the propagation phase. Soaking bulbs of lilies in 548 ppm dicofol for 30 minutes can significantly reduce mite numbers. Flooding soil with water for five days kills 96.1% of the mite infesting *Gladiolus* and all the mites are killed after 14 days. *R. robini* is known to be sensitive to sulphur, azocyclotin, cyhexatin, methidathion, phosmet, profenofos and prothiofos, as well as to hot and dry conditions.

References & suggested further reading. Ascerno *et al.* (1983); Afifi *et al.* (1987); Hassan *et al.* (1987); Chen (1990); Kassab and Hafez (1990); Conijn *et al.* (1997); Lesna *et al.* (1995, 1996, 2000).

8.4.2 *Rhizoglyphus echinopus* (Fumouze and Robin)

8.4.2.1 COMMON NAME

This species is known generally as the bulb mite.

8.4.2.2 DISTRIBUTION AND HOST PLANTS

This is a cosmopolitan species. It attacks bulbs and roots of many species, including *Freesia*, *Gladiolus*, hyacinth, lily, iris, *Narcissus* and tulips.

References & suggested further reading. Manson (1972); Fan and Zhang (2003).

8.4.2.3 APPEARANCE AND DAMAGE SYMPTOMS

The general appearance of *R. echinopus* is very similar to that of *R. robini*. Damage symptoms are also similar to those caused by *R. robini*. Damaged roots of *Freesia* and *Gladiolus* develop dark brown streaks and are often mined internally. Healthy corms grown in heavily infested soil will have distorted growing tips and leaves.

References & suggested further reading. Hussey *et al.* (1969); Alford (1994).

8.4.2.4 Diagnostic Characters

The adult female is 791-860 µm long. Dorsal idiosomal setae are relatively long; setae sc_1 are long (45-95 µm); the first two pairs of dorsomedian setae (c_1 and d_1 Fig. 8.1.2B) are longer than half of the distance between their bases. The supracoxal seta is thick, 45-50 µm long. The Grandjean's organ has a distinct forked tip. The bursa copulatrix has a large opening just posterior to the anal slit and opens internally into a large transverse sac with a V-shaped projection at each end. In the male, the aedeagus is broadly rounded with a short tube-like anterior opening.

References & suggested further reading. Manson (1972); Hughes (1976); Fan and Zhang (2003).

8.4.2.5 Life History and Biology

Development occurs above 9.4°C, which is lower than the threshold temperature for *R. robini* (11.8°C). The incubation period of eggs is shorter, but the longevity of adults and male:female sex ratio are greater at higher temperatures. Fecundity of *R. echinopus* is not affected by temperature (cf: temperature-dependent in *R. robini*; Fig. 8.5B). Developmental time and longevity of *E. echinopus* are slightly longer than those of *R. robini*, whereas the fecundity of *E. echinopus* is lower (Table 8.1).

References & suggested further reading. Sakurai *et al.* (1992); Ostovan and Kamali (1996).

Table 8.1 Life history parameters of *Rhizoglyphus robini* and *R. echinopus* at 27±1°C, feeding on potato tubes (data from Ostovan and Kamali, 1996).

	R. robini	*R. echinopus*
Egg to adult development time (days)	12.1	13.5
Pre-oviposition period (days)	1.0	0.9
Oviposition period (days)	22.5	31.5
Post-oviposition period (eggs)	8.2	13.4
Reproductive rate (eggs/day)	58	43
Fecundity (total number of eggs)	730 (450-870)	620 (400-850)

8.4.2.6 Control and Management

Control methods used against *R. robini* should be generally applicable to *R. echinopus*.

When released at relatively high predator:prey ratios, *Hypoaspis aculeifer* should provide effective control of *R. echinopus*. A female deutonymph of *Hypoaspis aculeifer* can consume 60 eggs, 132 larvae, 20 protonymphs, eight deutonymphs or four adults of *R. echinopus*. When feeding on the larvae of *R. echinopus*, an adult female of *H. aculeifer* can lay 2.8 eggs per day and a total of 114 eggs during her life.

Another laelapid mite, *Hypoaspis vacua*, can also develop successfully when feeding on nymphs of *R. echinopus*. Immature stages of *H. vacua* consume an average total of 33 nymphs of *R. echinopus*, whereas adult females consume 13 nymphs daily.

Rhizoglyphus echinopus is insensitive to many pesticides (at least nine pyrethroids, six organochlorines, four formamidines, 14 specific acaricides, diflubenzuron, nicotine and abamectin) but is known to be susceptible to dieldrin, endrin, aldrin, deltamethrin, chlorpyrifos, diazinon, azinphos-ethyl and carbofuran.

References & suggested further reading. Knowles *et al.* (1988); Ragusa and Zedan (1988); Abou-Awad *et al.* (1989); Das and Mishra (1995); Gencsoylu *et al.* (1998).

8.4.3 *Tyrophagus* species

Several species of *Tyrophagus* occasionally attack greenhouse vegetables and ornamentals. They are normally associated with decaying organic matter (e.g. straw bales) in the soil or other growing media. When mite density is high, a portion of the population may move onto plants and can cause injury to young leaves and flowering buds.

The symptoms of damage by different *Tyrophagus* species are quite similar on the same plant species but they vary greatly from plant species to plant species. On cucumber, melon, and pumpkin seedlings, mite feeding produces numerous small holes and yellowish spots on young leaves which later become deformed; leaves of mature cucumber may be skeletonized on the dorsal surface. On watermelon, tomato and *Capsicum* seedlings, infested leaves became lustrous, discoloured and deformed. On tomato, attacked seedlings become dwarfed and stunted. On *Gerbera*, mite feeding in the bud stage results in few fully expanded petals and malformation of the flower base. On begonia, mites damage the anther and cause discoloration to flowers; stems may be dwarfed and show corky streaks. On *Viola* and cyclamen, flowers become infertile due to mite damage to anthers. On kalanchoë, infected shoots show narrow brown streaks of corky tissue; mites can damage the growing tips and rudimentary leaves and can also induce the development of secondary buds, giving the plant a 'broomy' appearance. Light infestation of the flower bulbs produces feeding marks on the margins of the developing leaves, which may become deformed and develop streaks of corky spots.

8.4.3.1 *Tyrophagus putrescentiae* (Schrank)

Known as the mould mite, this species is a major pest of stored products and houses. It is known to attack cucumbers in greenhouses in the UK, Poland and Japan. It also occurs on ornamentals such as *Gerbera*, *Viola*, *Cymbidium* orchid, kalanchoë, *Freesia*, *Tulipa* and *Narcissus* in greenhouses in Europe.

This is a relatively small and slender mite (Fig. 8.2.5A), with adult females 320-420 μm long and adult males 280-350 μm long. The body surface is translucent and smooth, giving a shiny appearance. The appendages are also colourless. The eggs are oval and only slightly punctated on the surface. The adults have a spatulate supracoxal seta with long lateral setules (Fig. 8.3.6A). Setae c_1 and d_2 are subequal in length. Setae d_1 are 2.0-3.5 times as long as d_2. Tarsus I is shorter than the combined length of tibia I and genu I. Solenidion ω_1 on tarsus I is cylindrical. The aedeagus in the male is short and doubly-bent into an S-shape (Fig. 8.3.6A).

Life history traits of this species are influenced by temperature, relative humidity, photoperiod and types of food. Development occurs between 10 and 35°C and the optimum temperature for development and survival is around 30°C (Fig. 8.6), at which the life cycle is completed in 8.5 days feeding on brewer's yeast flake. The most favourable relative humidity is 85% to 95%. The fecundity and egg viability of this species decrease and the duration of life stages increases with increasing light exposure. On corms of *Freesia*, its population increases only 82 times during one generation (in comparison to 159 times on wheat germ). On *Freesia* corms, its fecundity reaches a maximum in the third week of life. It grows faster on corms of freesia and crocus than on either tulip or hyacinth bulb (Fig. 8.8). Soil provides a better substrate for colony development of *T. putrescentiae* than *Gerbera* flower heads.

Prevention of the infestation of the growing media and discouragement of the rapid growth of this species in the growing media can reduce its chance of moving onto plants. The soil or other growing media should be sterilized. Fungicides may be used to reduce fungal food for *T. putrescentiae* in the substrate. Soil predatory mites (e.g. *Hypoaspis* species) may be released to reduce mite buildup in the soil. Phytoseid predators such as *Neoseiulus cucumeris* and *N. barkeri* may be used against mites on leaves and flowers.

Parathion, applied as a spray or as dust, gives good control of *T. putrescentiae* on *Kalanchoë*, *Viola* or *Gerbera*, but dicofol, endosulfan, pirimiphos-methyl and cyhexatin are ineffective.

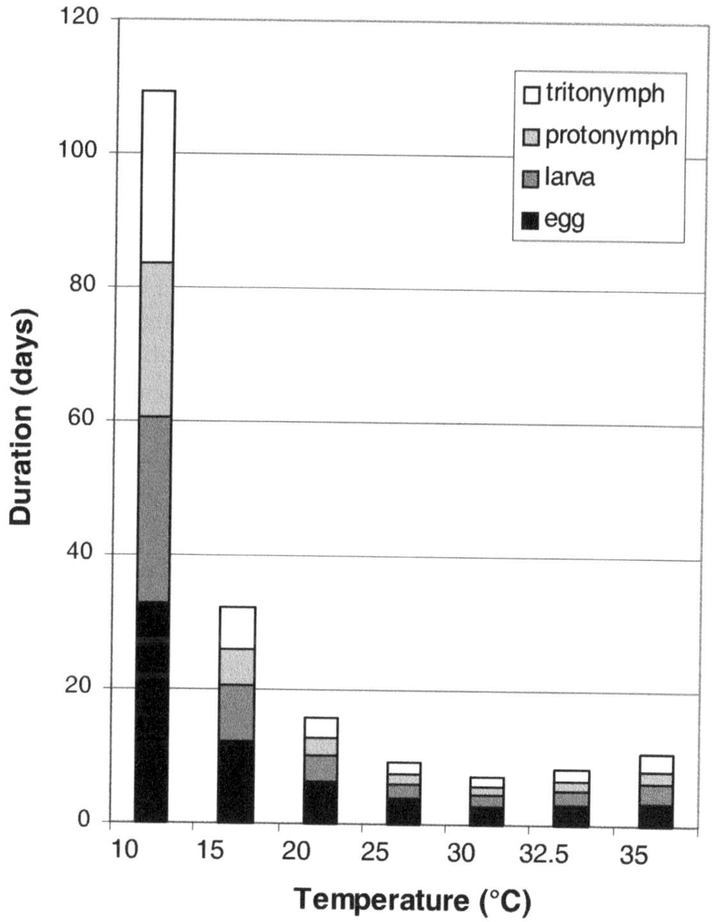

Fig. 8.6 *Tyrophagus putrescentiae*: Egg to adult developmental time in relation to temperature at relative humidity of 90 ± 5% (drawn from data in Sánchez-Ramos and Castañera, 2001).

References & suggested further reading. Griffiths and Southgate (1967); Chmielewski (1979); Czaikowska *et al.* (1988); Santos (1989); Czajkowska and Kropczynska (1991); Nakao (1991); Kohli and Mathur (1993); Li *et al.* (1998); Sanchez-Ramos and Castanera (2001).

8.4.3.2 *Tyrophagus similis* Volgin

This species is common on grasslands, in soil, in compost, in stored hay and houses, on mushrooms, and in nests of animals such as birds and insects. On greenhouse plants, it is known as the 'French fly' and has been found on leaves of cucumber, French beans, *Phlox* and spinach, stems of *Cineraria*, and bulbs of *Narcissus* in Europe. In Japan, *T. similis* attacks cucumbers, tomatoes, sweet peppers [*Capsicum*] and other crops. The most serious damage occurs to greenhouse grown spinach in early spring, with a positive correlation between the number of individuals of *T. similis* per plant and the percentage of plants damaged.

This is a relatively large mite, with adult females 600 μm long and adult males 500 μm long. The appendages are more tanned than those in *T. putrescentiae*. The eggs are oval and are distinctly punctated on the surface. The supracoxal seta is curved but not enlarged basally as in *T. putrescentiae*. Setae c_1, d_1 and d_2 are subequal in length (Fig. 8.4.7B). Solenidion ω_1 on tarsus I is enlarged distally. The aedeagus in the male is curved and truncated distally (Fig. 8.4.8B).

Life cycle of this species is temperature-dependent and is slightly faster than that of *T. putrescentiae* (Fig. 8.7). Development is shorter with more exposure to light. Egg to adult development can be completed in fewer than ten days at 25°C reared on wheat germ (Fig. 8.7). At 25°C, mite fecundity is much lower than at 10°C (663 eggs/female). On cucumber leaves, life cycle is completed in 17-24 days at 22°C. Adults live for up to three weeks and lay 40-60 eggs. *T. similis* females show limited survival at 35°C and higher temperatures, and at 53% and 66% RH.

Methods for control of this mite are similar to those for *T. putrescentiae* (see 8.4.3.1).

On cucumber and French beans, early infestation may be treated by 'smoking' with parathion. When there is a high population of mites in straw bales, spraying both the plants and the surface of substrate using 0.01% parathion may be necessary. Severe infestations may be controlled by spraying dicofol at 0.01%. On spinach, the mite can be controlled by benzoximate, endosulfan, fenpropathrin and benzoximate, but these chemicals do not kill the eggs and are toxic to predaceous mites.

References & suggested further reading. Hussey *et al.* (1969); Hughes (1976); Czaikowska *et al.* (1988); Ippolito and Triggiani (1988); Nakao (1989, 1991); Al-Safadi (1990, 1991); Kasuga and Amano (2000).

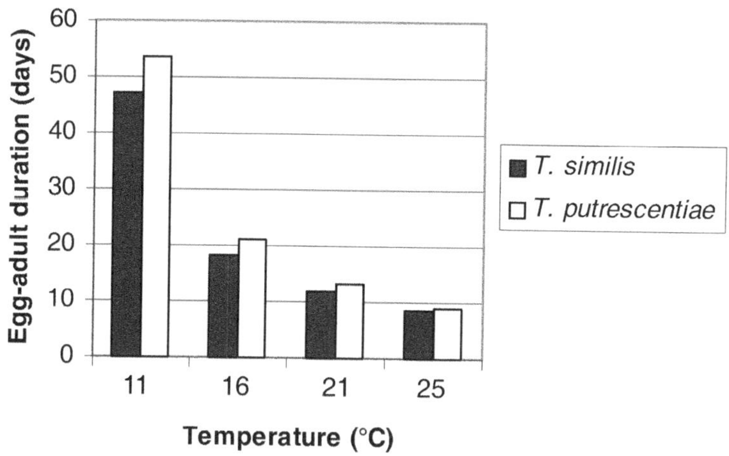

Fig. 8.7 Egg to adult developmental time of *Tyrophagus similis* and *Tyrophagus putrescentiae* in relation to temperature reared in wheat germ (drawn from data in Al-Safadi, 1991).

8.4.3.3 *Tyrophagus longior* (Gervais)

This cosmopolitan species is generally a pest of stored products and houses. It is an occasional pest of greenhouse cucumber in Europe and causes damage to the crop. In the UK, it is also found on *Verbena*, *Lavatera* and *Delphinium ajacis* [*Consolida ambigua*], although no organic matter such as manure or straw is associated with either infestation. In Italy, it attacks orchids of the genus *Cymbidium* cultivated in greenhouses. The most heavily infested orchid species is *Cymbidium clauboda*. The mite vectors fungal spores that prevent the flowers from opening. Damage is slight or negligible on other species of *Cymbidium*.

This is a relatively large mite, with adult females 530-670 μm long and adult males 330-535 μm long. The appendages are more tanned than in *T. putrescentiae*. The eggs are distinctly punctated on the surface. The supracoxal seta is curved but not enlarged basally as in *T. putrescentiae*. Setae c_1 and d_2 are subequal in length. Setae d_1 are 1.0-1.3 times as long as d_2 (Fig. 8.3.5B). Tarsus IV is longer than combined length of tibia IV and genu IV. Solenidion ω_1 on tarsus I is relatively long and gradually tapered slightly (Fig. 8.4.8A). The aedeagus in the male tapers toward its free end and looks like the spout of a teapot (Fig. 8.4.8A).

The life cycle is completed in two to three weeks at 23°C and 87% RH. Reared on grains of wheat at 20°C in test tubes, its population can increase from five pairs to 2,200 mites before population crash when RH is 90%. At 75% RH, its population peaks at 1,000 mites only and then decreases. Little is known about its development and reproduction on leaves or flowers of cucumber and ornamental plants that it attacks.

Methods for control of this mite should be similar to those for *T. putrescentiae* (see 8.4.3.1).

On *Verbena, Lavatera* and *D. ajacis* [*C. ambigua*], this mite can be effectively controlled by a high-volume spray of dichlorvos. On *Cymbidium*, effective methods include preventive applications of fungicides, sterilization of the compost and cultivation of species resistant to *T. longior*. On cucumber, this mite can be controlled by sprays of pirimiphos-methyl (25g/100 litres).

References & suggested further reading. Hughes (1976); Chmielewski (1979, 1984); Ciampolini *et al.* (1985); Hussey (1985); Czaikowska *et al.* (1988); Buxton (1989); Parkinson (1990); Voigt (1990).

8.4.3.4 *Tyrophagus neiswanderi* Johnston & Bruce

This species was originally found on glasshouse cucumber plants in Ohio, USA, although it was subsequently found in stored products and nests of animals. In greenhouses in Europe, it is found on leaves of cucumbers, growing tips of chrysanthemum cuttings, flowers of *Gerbera* and cyclamen, bulbs of *Narcissus, Tulipa* and *Hippeastrum*, and corms of *Freesia*. It also attacks *Cymbidium* orchids in New Zealand and *Phalaenopsis* orchids, cucumbers and other plants in greenhouses in Japan.

This is a relatively large mite, with adult females 410-550 μm and adult males 380-460 μm long. The eggs are distinctly punctated on the surface. The supracoxal seta is enlarged basally as in *T. putrescentiae*, but not as elongate distally as in the latter. Setae c_1 and d_2 are subequal in length. Setae d_1 are 1.4-1.7 times as long as d_2 (Fig. 8.4.7A). Tarsus IV is shorter than the combined length of tibia IV and genu IV. Solenidion ω_1 on tarsus I is cylindrical and curved slightly. The aedeagus in the male is curved twice but not tapered towards the distal end.

This species develops a bit more slowly than *T. putrescentiae* and the egg to adult development is completed in two to three weeks at 25°C (Fig. 8.8). It grows faster on corms of freesia and crocus than on either tulip or hyacinth bulbs.

Methods for control of this mite should be similar to those for *T. putrescentiae* (see 8.4.3.1).

References & suggested further reading. Johnston and Bruce (1965); Griffith and Southgate (1967); Hughes (1976); Chmielewski (1979, 1984); Martin and Workman (1985); Czaikowska *et al.* (1988); Czajkowska and Kropczynska (1991); Kurosa and Nakao (1993); Fischer (1993); Kadono and Endo (1996).

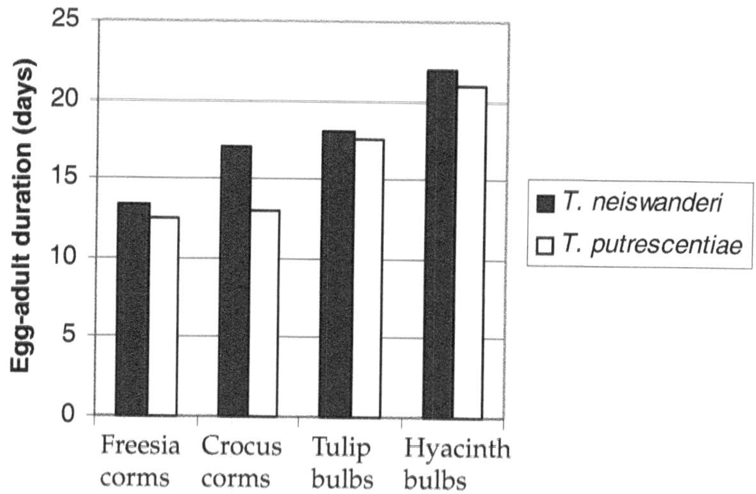

Fig. 8.8 Egg to adult developmental time of *Tyrophagus neiswanderi* and *Tyrophagus putrescentiae* on four host plants at 25°C (drawn from data in Czajkowska and Kropczynska, 1991).

8.4.3.5 *Tyrophagus perniciosus* Zachvatkin

This mite is usually associated with stored products and houses. In Japan, *T. perniciosus* attacks cucumbers, melons and pumpkins. It is known to feed on the mycelium of *Sporidesmium mucosum* var. *pluriseptatum*, the fungus causing brown mosaic disease of cucumber in greenhouses in the former Kazakh SSR, USSR.

This is a relatively large mite, with adult females 550-700 µm long and adult males 450-500 µm long. The supracoxal seta gradually expands basally with setules becoming shorter towards the distal end (Fig. 8.3.6B). Setae d_1 are 2.5-4.5 times as long as d_2. Solenidion ω_1 on tarsus I is relatively short and slightly expanded distally. The aedeagus in the male is arched and truncate distally (Fig. 8.3.6B).

Little is known about the biology and control of this species.

References & suggested further reading. Hughes (1976); Sadieva (1984); Nakao (1991).

8.4.3.6 *Tyrophagus curvipenis* Fain & Fauvel

This species is known from orchids in a greenhouse in Portugal. It feeds on algae covering the wooden structures of the greenhouse and occasionally enters flowers where they feed on pollen. Nothing is known about its economic significance, biology and control.

References & suggested further reading. Fain and Fauvel (1993).

8.4.4 *Mycetoglyphus fungivorus* (Oudemans)

This species is found in mushrooms, celery waste, lettuce, decaying radishes, stored hay and straw, the nests of animals and on grassland. In Japan, it attacks spinach grown in greenhouses. The damage caused by *M. fungivorus* is similar to that caused by *T. similis*. The most serious damage occurs to greenhouse cultivated spinach in early spring.

This is a relatively large mite, with adult females measuring 500-600 μm long and adult males 400-600 μm long. It looks very similar to *T. similis*, with tanned appendages. External vertical setae v_1 are less than a quarter as long as internal vertical setae v_2 (Fig. 8.2.4A). The supracoxal seta is slender and curved, with minute setules. Setae d_1 are 1.5-2.0 times as long as d_2. Solenidion ω_1 on tarsus I is rod-like. The aedeagus in the male is a long, curved tube.

Little is known about the biology and control of this mite. Effective methods against *Tyrophagus* should be applicable to this species.

References & suggested further reading. Hughes (1976); Nakao (1989).

8.5 References cited

Abou-Awad, B.A., Nasr, A.K., Gomaa, E.A. and Abou-Elela, M.M. (1989) Feeding, development and reproduction of the predatory mite, *Hypoaspis vacua* on various kinds of food substances (Acari: Laelapidae). *Insect Science and its Application* 10, 503-506.

Afifi, A.M., Hassan, M.F. and Nawar, M.S. (1987) Notes on the biology feeding habits of *Protogamasellus minutus* Hafez, El-Badry and Nasr (Acari: Gamasida: Ascidae). *Bulletin de la Societe Entomologique d'Egypte* 66, 251-259.

Alford, D.V. (1994) *A Colour Atlas of Pests of Ornamental Trees, Shrubs and Flowers.* Wolfe Publishing Ltd., London. 448 pp.

Al-Safadi, M.M. (1990) The effect of light on the life cycle of the Acari *Tyrophagus similis* (Astigmata: Acaridae). *Revue d'Ecologie et de Biologie du Sol* 27, 459-465.

Al-Safadi, M.M. (1991) The effect of temperature on the developmental rate of the *Tyrophagus similis* Volgin (Acari: Acaridae). *Revue d'Ecologie et de Biologie du Sol* 28, 341-347.

Ascerno, M.E., Pfleger, F.L., Morgan, F. and Wilkins, H.F. (1983) Relationship of *Rhyzoglyphus robini* (Acari: Acaridae) to root rot control in greenhouse-forced Easter lilies. *Environmental Entomology* 12, 422-425.

Buxton, J.H. (1989) *Tyrophagus longior* (Gervais) (Acarina; Acaridae) as a pest of ornamentals grown under protection. *Plant Pathology* 38, 447-448.

Chen, J.S. (1990) The biology, insecticide resistance, and control countermeasures of bulb mites. *Chinese Journal of Entomology Special Publication* 3, 93-106.

Chmielewski, W. (1979) Mites occurring on greenhouse cucumbers. *Ochrony Roslin* 10, 5-7.

Chmielewski, W. (1984) *Tyrophagus longior* (Gerv., 1844) (Acarina, Acaridae)-bioecology, occurrence and harmfulness. *Ochrony Roslin* 26, 69-87.

Ciampolini, M., Lugaresi, C., Rota, P.A. and Cappella, A. (1985) A mite injurious to orchids in the greenhouse. *Informatore Agrario* 41, 33, 29-33.

Conijn, C.G.M., Lesna, I. and Altena, K. (1997) Biological control of the bulb mite *Rhizoglyphus robini* by the predatory mite *Hypoaspis aculeifer* on lilies: implementation in practice. *Acta Horticulturae* 430, 619-624.

Czajkowska, B. and Kropczynska, D. (1991) The influence of different host plants on the reproductive potential of *Tyrophagus putrescentiae* (Schrank) and Tyrophagus neiswanderi Johnston and Bruce (Acaridae). In: Schuster, R. and Murphy, P.W. (eds) *The Acari. Reproduction, Development and Life History Strategies*. Chapman & Hall, London, UK, pp. 313-317.

Czaikowska, B., Vrie, M. van de and Kropczynska, D. (1988) Mites of the genus *Tyrophagus* as pests of ornamentals in greenhouses. *Mededelingen van de Faculteit Landbouwweten-schappen, Rijksuniversiteit Gent* 53(2b), 799-809.

Das, T.K. and Mishra, D.N. (1995) Toxicity of some synthetic pyrethroids to *Rhizoglyphus echinopus* (Acarina: Acaridae), a pest of mushroom. *Environment and Ecology* 13, 366-368.

Diaz, A., Okabe, K., Eckenrode, C.J., Villani, M.G. and O'Connor, B.M. (2000) Biology, ecology, and management of the bulb mites of the genus *Rhizoglyphus* (Acari: Acaridae). *Experimental & Applied Acarology* 24, 85-113.

Fain, A. and Fauvel, G. (1993) *Tyrophagus curvipenis* n. sp. from an orchid cultivation in a greenhouse in Portugal (Acari: Acaridae). *International Journal of Acarology* 19, 95-100.

Fan, Q.-H. and Zhang, Z.-Q. (2003) *Rhizoglyphus echinopus* and *Rhizoglyphus robini* (Acari: Acaridae) from Australia and New Zealand: identification, host plants and geographical distribution. *Systematic & Applied Acarology Special Publications* 16, 1-16.

Fischer, S. (1993) Observation of a new pest of cucumber in western Switzerland, *Tyrophagus neiswanderi* Johnston & Bruce (Acari, Acaridae). *Revue Suisse de Viticulture, d'Arboriculture et d'Horticulture* 25(2), 103-104.

Gencsoylu, I., Liu, W., Usmani, K.A. and Knowles, C.O. (1998) Toxicity of acaricides to the bulb mite *Rhizoglyphus echinopus* (Acari: Acaridae). *Experimental and Applied Acarology* 22, 343-351.

Gerson, U., Capua, S. and Thorens, D. (1983) Life history and life tables of *Rhizoglyphus robini* Claparede (Acari: Astigmata: Acaridae). *Acarologia* 24, 439-448.

Griffiths, D.A. and Southgate, B.J. (1967) Mites of the genus *Tyrophagus* as pests of cucumber. In: *Pest Infestation Research 1967*. Agricultural Research Council, 3.

Hassan, M.F., Afifi, A.M. and Nawar, M.S. (1987) *Dendrolaelaps sayedi*, a new digamasellid mite (Acari: Gamasida) from Egypt, with notes on its biology. *Bulletin de la Societe Entomologique d'Egypte* 66, 227-235.

Hughes, A.M. (1976) *Mites of Stored Food and Houses. Technical Bulletin of the Ministry of Agriculture, Fisheries and Food, No. 9*. HMSO, London, 400 pp.

Hussey, N.W. (1985) Cucumbers. In: Hussey, N.W. and Scope, N.E.A. (1985) *Biological Pest Control. The Glasshouse Experience*. Blandford Press, Dorset, pp. 175-179.

Hussey, N.W., Read, W.H. and Hesling, J.J. (1969) *The Pests of Protected Cultivation: The Biology and Control of Glasshouse and Mushroom Pests*. Edward Arnold (Publishers) Ltd., London. 404 pp.

Ippolito, R. and Triggiani, O. (1988) Experiments on the efficacy of some acaricides against *Tyrophagus similis* Volgin (Acaridida - Acaridae) on spinach. *Informatore Fitopatologico* 1988, 38, 43-48.

Johnston, D.E. and Bruce, W.A. (1965) *Tyrophagus neiswanderi*, a new acarid mite of agricultural importance. *Research Bulletin of Ohio Agricutural Experiment Station* 977, 1-17.

Kadono, F. and Endo, M (1996) Seasonal population trend and control of *Tyrophagus neiswanderi* Johnston et Bruce on Phalaenopsis in the greenhouse. *Proceedings of the Kanto Tosan Plant Protection Society* 43, 247-252.

Kassab, A.S. and Hafez, S.M. (1990) Use of powdered sulfur against the bulb mite, *Rhizoglyphus robini*, and its effect on nematodes in garlic field soil. *Annals of Agricultural Science Cairo* 35, 533-541.

Kasuga, S. and Amano, H (2000) Influence of temperature on the life history parameters of *Tyrophagus similis* Volgin (Acari: Acaridae). *Applied Entomology and Zoology* 35, 237-244.

Knowles, C.O., Errampalli, D.D. and El-Sayed, G.N. (1988) Comparative toxicities of selected pesticides to bulb mite (Acari: Acaridae) and twospotted spider mite (Acari: Tetranychidae). *Journal of Economic Entomology* 81, 1586-1591.

Kohli, R. and Mathur, S. (1993) Life processes of *Tyrophagus putrescentiae* (Schrank) (Acarina: Acaridae) as influenced by photoperiod. *Crop Research Hisar* 6, 311-316.

Kurosa, K. and Nakao, H. (1993) *Tyrophagus neiswanderi* Johnston *et* Bruce. In: Ehara, S. (ed.) *Plant Mites of Japan in Colors*. Zenkoku Noson Kyoiku Kyokai, Tokyo, pp. 156-157.

Lesna, I., Sabelis, M.W., Bolland, H.R. and Conijn, C.G.M. (1995) Candidate natural enemies for control of *Rhizoglyphus robini* Claparede (Acari: Astigmata) in lily bulbs: exploration in the field and pre-selection in the laboratory. *Experimental & Applied Acarology* 19, 655-669.

Lesna, I., Sabelis, M.W. and Conijn, C. (1996) Biological control of the bulb mite, *Rhizoglyphus robini*, by the predatory mite, *Hypoaspis aculeifer*, on lilies:

predator-prey interactions at various spatial scales. *Journal of Applied Ecology* 33, 369-376.

Lesna, I., Conijn, C.G.M., Sabelis, M.W. and Straalen, N.M. van (2000) Biological control of the bulb mite, *Rhizoglyphus robini*, by the predatory mite, *Hypoaspis aculeifer*, on lilies: predator-prey dynamics in the soil, under greenhouse and field conditions. *Biocontrol Science and Technology* 10, 179-193.

Li, L.S., Chen, B., Xia, J. and Zhang, X.W. (1998) Influence of temperature and controlled atmosphere on development and reproduction of the mold mite, *Tyrophagus putrescentiae* (Acari: Acaridae). *Systematic and Applied Acarology* 3, 113-120.

Liu, T.S. and Tzeng, A.K. (1994) Biological and damaging characteristics of bulb mite, *Rhizoglyphus robini* Claparede. *Plant Protection Bulletin Taipei* 36, 177-187.

Manson, D.C.M. (1972) A contribution to the study of the genus *Rhizoglyphus* Claparède, 1869 (Acarina: Acaridae). *Acarologia* 13, 621-650.

Martin, N.A. and Workman, P. (1985) Pest control in boxes of Cymbidium orchid flowers with dichlorvos-impregnated plastic. *Proceedings of New Zealand Pest Control Conference* 38, 169-171.

Meyer, M.K.P. (Smith) (1981) Mite pests of crops in Southern Africa. *Science Bulletin Department of Agriculture and Fisheries Republic of South Africa* 397, 1-92.

Nakao, H. (1989) Studies on acarid mites injurious to vegetable plants (Acari: Astigmata) I. Occurrence of damage to spinach by acarid mites. *Bulletin of Hokkaido Prefectural Agricultural Experiment Stations* 59, 41-47.

Nakao, H. (1991) Studies on acarid mites (Acari: Astigmata) damaging vegetable plants. II. Damage to vegetable seedlings. *Japanese Journal of Applied Entomology and Zoology* 35, 303-309.

O'Connor, B.M. (1982) Acari: Astigmata. In: Parker, S. (ed.) *Synopsis and Classification of Living Organisms*, Vol. 2. McGraw-Hill, New York, pp. 146-169.

Okabe, K. and Amano, H. (1991) Penetration and population growth of the robine bulb mite, *Rhizoglyphus robini* Claparede (Acari: Acaridae), on healthy and Fusarium-infected rakkyo bulbs. *Applied Entomology and Zoology* 26, 129-136.

Ostovan, H. and Kamali, K. (1996) Biology of two bulb mites: *Rhizoglyphus echinopus* (Fumouze and Robin) and *R. robini* Claparede under two different laboratory conditions. *Journal of Agricultural Sciences Islamic Azad University* 2(5-6), 47-59.

Parkinson, C.L. (1990) Population increase and damage by three species of mites on wheat at 20°C and two humidities. *Experimental & Applied Acarology* 8, 179-193.

Ragusa, S. and Zedan, M.A (1988) Biology and predation of *Hypoaspis aculeifer* (Canestrini) (Parasitiformes, Dermanyssidae) on *Rhizoglyphus echinopus* (Fum. & Rob.) (Acariformes, Acaridae). *Redia* 76, 213-126.

Raut, S.K. and Sarkar, R. (1991) The influence of temperature on the life-cycle of *Rhizoglyphus robini* Claparede (Acari: Acaridae). *International Journal of Acarology* 17, 145-148.

Sadieva, B.E. (1984) A mite and the pathogen of brown mosaic. *Zashchita Rasteni* 1984(7), 20.

Sakurai, H., Inaba, T. and Takeda, S. (1992) Effect of temperature on the development of bulb mite, *Rhizoglyphus echinopus*. *Research Bulletin of the Faculty of Agriculture, Gifu University* 1992(57), 81-90.

Sánchez-Ramos, I. and Castañera, P. (2001) Development and survival of *Tyrophagus putrescentiae* (Acari: Acaridae) at constant temperatures. *Environmental Entomology* 30, 1082-1089.

Santos, M.A. (1989) Mortality and fecundity of *Tyrophagus putrescentiae* (Acari: Acaridae) in relation to population density. *Environmental Entomology* 18, 1060-1065.

Voigt, P. (1990) Damage caused by mould mites in cucumbers. *Nachrichtenblatt Pflanzenschutz* 44, 246-247.

Wang, C.L. and Lin, R.T. (1986) Studies on soil treatments for the control of the bulb mite (*Rhizoglyphus robini* Claparede) on gladiolus. *Journal of Agricultural Research of China* 35, 230-234.

Other Pest Mites 9

9.1 Introduction

Major groups of pest mites of significance on greenhouse plants have been dealt with in the last five chapters. A minor family, Siteroptidae, has only one species of importance in greenhouses. There are a few other families of phytophagous mites, which occasionally appear in greenhouses or have not been reported as pests in greenhouses, but could potentially occur in greenhouses. These are introduced here in this chapter.

9.2 Siteroptidae

The family Siteroptidae was previously placed in the family Pygmephoridae in the order Prostigmata, but many acarologists now consider it as a separate family. Both Siteroptidae and Pygmephoridae belong to the superfamily of Pygmephoroidea, which is closely related to Tarsonemoidea (see Chapter 6).

Pygmephoroids are small mites, with most species ranging from 200 to 300 μm. The chelicerae and subcapitulum are fused to the gnathosomal capsule, which has a pair of small cheliceral stylets and reduced palps. The prodorsal sclerite of the female does not cover the entire propodosoma, which has two or three pairs of dorsal setae, a pair of anterior stigmata and associated peritremes, and a pair of lateral capitate trichobothria (Fig. 9.1A). The bases of legs I-II are twice as widely apart as the bases of legs III-IV. The legs are four- or five-segmented. Trochanter IV of the female is quadrate or rectangular. Femur I has three to five setae; if three setae are present, the dorsal setae are highly modified, not setiform. Leg apoteles I have one claw, rarely two claws, with or without a pulvillar empodium; apoteles II-IV have two claws and a pulvillar empodium; apoteles IV sometimes lack claws.

Many pygmephoroid species are known to be fungivorous. Some *Siteroptes* species are pests of grasses and cereals and are responsible for transmissions of pathogenic fungi. Females of some genera are known to be physogastric and have a specialized form (phoretomorph) that is adapted to attach to insects for dispersal. Physogastric females can give birth to over 100 offspring (mostly males and females, but also some larvae).

Only one species, *Siteroptes avenae* Müller [=*S. graminum* (Reuter), not *S. cerealium* Kirchner], is of economic importance in greenhouses. It is normally a pest of grasses and cereals, but it is known from carnations in the UK and mainland Europe, occurring in association with the fungus *Fusarium poae*, which causes carnation bud rot. This mite feeds on fungal spores and is not phytophagous. It damages carnation by vectoring the fungus; spores are carried by the adult mite, on the body or inside the spirothecae (Fig. 9.1B). Infected buds may appear normal for 16 days, but by the time the calyx appears abnormal, all the petals and sepals have rotted.

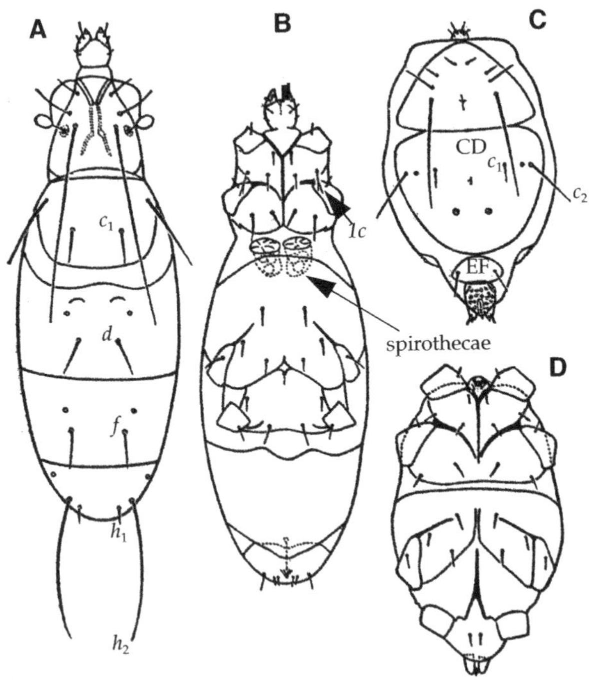

Fig. 9.1 *Siteroptes avenae* Müller. A, adult female, dorsal view; B, adult female, ventral view; C, adult male, dorsal view; D, adult male, ventral view (after Su and Ding, 1984).

Adult females are elliptical, 240-270 μm (Fig. 9.1A), and yellowish in colour when alive. Dorsal setae c_1, d and f are subequal in length. Setae h_2 is about nine times as long as h_1. Coxal setae $1c$ are forked distally (Fig. 9.1.B). A pair of spirothecae is located behind coxae II.

Adult males are ovoid, but tapered posteriorly (Fig. 9.1C). Setae c_2 are about three times as long as c_1. Prodorsal plate and plate CD have a median apodeme. Plate EF is small and semicircular.

The life cycle consists of egg, larva, nymph and adult stages. Reproduction is parthenogenetic and immature stages develop inside the adult female. Mating also occurs inside the mother's body. Unmated females give birth to males. The hysterosoma of the female is swollen and may increase to 500 times of the original body size.

Infestation of carnations in greenhouses often occurs in the autumn following a spell of hot and dry weather.

There are no published reports on control measures for this mite on carnations, but the control of the fungi will be the key to the management of the problem. Use common sense control methods when appropriate (Table 1.5).

References & recommended further reading. Cooper (1937, 1940); Hussey *et al.* (1969); Su and Ding (1984); Suski (1984); Alford (1994).

9.3 Penthaleidae

The Penthaleidae is a small family of the Eupodoidea belonging to the order Prostigmata. There are five genera, two of which have a single species each and are of agricultural importance.

These are medium-sized, weakly sclerotized mites. The cheliceral bases are separate. Two pairs of setae are present on the ventral surface of the gnathosoma. The chilicerae are chelate, each bearing a single seta. The movable digits are shear-like and smooth, whereas the fixed digits are slender and are often finger-like distally. The palps are four-segmented, without a tibial claw. The stigmata are located between the bases of the chelicerae, but external peritremes are absent. The idiosoma is unornamented, with moderate to large numbers of setae. The anal opening is dorsal or dorsoterminal. Tarsus II has a recumbent solenidion. Leg apoteles have paired hooked claws and ciliated pad-like empodium.

The life cycle of the Penthaleidae is typical of the Prostigmata: immatures consist of a larva and three nymphal stages. Sperm transfer is by deposited spermatophores.

The red-legged earth mite, *Halotydeus destructor* (Tucker), is an important pest of many low-growing crops, especially annual, broad-leaved plants and grasses, in Australia and South Africa. It is occasionally found feeding on seedlings of greenhouse-grown vegetables (peas and tomato) and ornamental annuals. *Penthaleus major* (Duges) infests cereals, grasses, vegetables and ornamental flowers in temperate and subtropical regions. This species may also occur on ornamentals grown in greenhouses.

References & suggested further reading. Jeppson *et al.* (1975); Kethley (1990); Qin (1996).

9.4 Tydeidae

The Tydeidae is a family of the Tydeoidea belonging to the order Prostigmata. There are over 300 species in more than 40 genera.

Tydeids are small, soft-body mites. The chelicerae are fused together, with styliform movable digits, but without fixed digits. The ventral surface of gnathosoma has two pairs of subcapitular setae and two pairs of adoral setae. The palps are four-segmented, usually of a very characteristic appearance with a very short third segment between much longer segments. The stigmata is located at bases of chelicerae. Idiosomal cuticles are striated, reticulate or verrucose. The sejugal furrow is present. The prodorsum has a pair of trichobothria and three pairs of normal setae. The hysterosoma has up to 12 pairs of dorsal setae. Legs terminate in paired true claws and a pad-like empodium (but claws and empodium on leg I are absent in some genera). Femur III is entire, but femur IV may be divided.

Most species are fungivorous, some are predatory and some are facultatively phytophagus. *Tydeus* [=*Orthotydeus*] *californicus*, *Tydeus caudatus* Duges, *Tydeus praefatus* and *Lorryia formosa* are known to cause significant damage to plants by feeding on leaves.

References & suggested further reading. Jeppson *et al.* (1975); André (1980); Bozai (1997).

9.5 Tuckerellidae

This family belongs to the superfamily Tetranychoidea in the Prostigmata. They are known as the peacock spider mites (peacock mites) because of the elaborate ornamentations on the dorsal surface of their bodies. There is only a single genus *Tuckerella*, with over 20 species known in the world.

The peacock mites are similar to spider mites in size and most species are red to orange with white setae. The chelicerae are fused together into a stylophore as in spider mites. The movable digits are styliform and recurved basally. The palps are four-segmented with a long tibial claw. The stigmata is located at the base of the stylophore. The idiosoma dorsally has a prodorsal shield bearing four pairs of palmate setae and three successive hysterosomal shields bearing 18-20 pairs of palmate setae. Dorsal shields are strongly reticulate or with raised lines. The prodorsum has two pairs of eyes. The idiosoma has a caudal row of about eight pairs of setae of which five to seven pairs are very long (about as long as the idiosoma). All legs are five-segmented, with true claws bearing tenent hairs.

The life cycle consists of the egg, larva, protonymph, deutonymph, tritonymph and adult stages. The developmental time of the Tuckerellidae is longer than that of spider mites, which have only two nymphal stages. One generation per year is known in some peacock mites.

All species are phytophagous and have been found on a wide range of host pants, which includes some crops and ornamental plants. These mites have not been reported as important pest species and little is known about how to control them. On plants that do not tolerate low pest densities (e.g. some ornamental flowers), use common sense control methods when appropriate (Table 1.5).

References & suggested further reading. Ochoa (1989); Kethley (1990).

9.6 Oribatida

The oribatid mites are the most common inhabitants of soil and litter. However, some oribatid species are associated with plants and have been found in greenhouses, with unknown economic significance.

Hemileius clavatus Aoki of the family Scheloribatidae is found on the leaves and stems of greenhouse-grown *Cymbidium* in Japan. Another species, *Hemileius thujae* Choi & Cho, is found on Royal Azalea (*Rhododendron schlippenbachii*) and other ornamental and landscape plants in Korea.

Mochlozetes penetrabilis Grandjean of the family Mochlozetidae is known from leaves and stems of greenhouse-grown *Cymbidium* in Japan.

Perlohmannia dissimilis (Hewitt) of the family Perlohmanniidae is known to cause damage to the root systems of potato, strawberry and tulip and could be found on strawberries and tulips in greenhouses.

References & suggested further reading. Jeppson *et al.* (1975); Aoki (1993a,b); Choi and Cho (1995).

9.7 References cited

Alford, D.V. (1994) *A Colour Atlas of Pests of Ornamental Trees, Shrubs and Flowers.* Wolfe Publishing Ltd., London. 448 pp.

André, H. (1980) A generic revision of the family Tydeidae (Acari: Actinedida) IV. Generic descriptions, keys and conclusions. *Bulletin et annales de la Société royale entomologique de Belgique* 116,103-168.

Aoki, J. (1993a) *Hemileius clavatus* Aoki. In: Ehara, S. (ed.) *Plant Mites of Japan in Colors.* Zenkoku Noson Kyoiku Kyokai, Tokyo, pp 156-157.

Aoki, J. (1993b) *Mochlozetes penetrabilis* Grandjean. In: Ehara, S. (ed.) *Plant Mites of Japan in Colors.* Zenkoku Noson Kyoiku Kyokai, Tokyo, pp 170-171.

Bozai, J. (1997) Data on the fauna of Tydeidae of Hungary (Acari, Tydeidae). *Novenyvedelem* 33(2), 77-79.

Choi, S.S. and Cho, M.R. (1995) A new species of oribatid mite (Acari: Oribatida) collected from landscape plants. *Korean Journal of Entomology* 25(3), 197-199.

Cooper, H.L. (1937) Reproductive behavior and haploid parthonogenensis in the grass mite, *Pediculopsis graminum* (Reut.). *Proceedings of the National Academy of Sciences, Washington* 23, 41-44.

Cooper, H.L. (1940) Relation of *Pediculopsis graminum* and *Fusarium poae* to cental rot of carnations. *Phytophathology* 30, 853-859.

Hussey, N.W., Read, W.H. and Hesling, J.J. (1969) *The Pests of Protected Cultivation: The Biology and Control of Glasshouse and Mushroom Pests.* Edward Arnold (Publishers) Ltd., London. 404 pp.

Jeppson, L.R., Baker, E.W. and Keifer, H.H. (1975) *Mites Injurious to Economic Plants.* University of California Press, Berkeley, California. 614 pp.

Kethley, J. (1990) Acarina: Prostigmata (Actinedida). In: Dindal, D.L. (ed.) *Soil Biology Guide.* John Wiley & Sons, New York, pp. 667-756.

Ochoa, R. (1989) The genus *Tuckerella* in Costa Rica (Acari: Tuckerellidae). *International Journal of Acarology* 15, 205-207.

Qin, T.K. (1996) A review and cladistic analysis of the Eupodoidea (Acari: Acariformes). *Systematic and Applied Acarology* 1, 77-105.

Su, D.-S. and Ding, T.-Z. (1984) Tarsonemina. In: Ma, E.P. (ed.) *Agricultural Acari in China [Zhonggu Nongye Manlei].* Shanghai Science & Technology Press, Shanghai, China, pp. 226-253 [in Chinese].

Suski, Z.W. (1984) On the identity of pyemotid mites associated with the silver-top disease of grasses. In: Griffith, D.A. and Bowman, C.E. (eds) *Acarology VI.* Vol. 1. Ellis Horwood Limited, Chichester, UK, pp. 174-179.

Part III
Beneficial Mites

Identification, biology and role in biological control of predatory mites

- Phytoseiid mites
- Laelapid mites
- Other predatory mites

Phytoseiid Mites

10

10.1 Introduction

Phytoseiid mites belong to the family Phytoseiidae of the order Mesostigmata. Phytoseiid mites are predators of spider mites and other small mites and insects on plants. Some species also feed on nematodes, fungal spores, pollen and exudates from plants, but rarely plant tissue. Several members of this family are of great importance in the biological control of spider mites and thrips in greenhouse crop production.

The Phytoseiidae is a large family of worldwide distribution. More than 1,600 species belonging to over 70 genera are known in the world. The family consists of three subfamilies: Amblyseiinae, Phytoseiinae and Typhlodrominae. Effective biocontrol agents occur in all three subfamilies, although most commercially available species that are commonly used in greenhouses belong to the genera *Neoseiulus* and *Phytoseiulus* in the Amblyseiinae.

There has been great confusion about the use of generic names of phytoseiid mites due to the different generic classifications of the family proposed by different specialists. For example, one of the most commonly used acarine biocontrol agent of thrips and other small mites in greenhouses has been known as *Amblyseius cucumeris* and *Neoseiulus cucumeris*. In this book, we adopt names that are commonly accepted by most phytoseiid specialists.

References & recommended further reading. Chant (1985, 1992); Helle and Sabelis (1985); Moraes *et al.* (1986); Chant and McMurtry (1994).

10.2 Morphological characters

Phytoseiids are medium-sized mites (mostly 250-400 μm) with relatively long legs. Many species are pale, yellowish, orange or dark brown.

The idiosoma is covered with a single entire shield (rarely two shields) with no more than 24 pairs of setae (dorsal setae J_1, J_3 and J_4 are absent). The stigmata open ventrolaterally at the level between the coxae of leg III and leg IV (Fig. 2.9.1A).

The sternal shield of the female has three pairs of setae. A pair of metasternal shields is small, each bearing one seta. The female genital shield is truncated posteriorly (Fig. 2.9.1A). The male genital opening is on the anterior margin of the sternal shield.

The male chelicera bears a spermatodactyl (Fig. 2.2B), the shape of which has diagnostic value.

Leg I terminates in well-developed ambulacra (Fig. 2.6B). Tibia I has five dorsal and two or three ventral setae. The distal leg segments often have one elongate or differently-shaped macroseta.

References & recommended further reading. Chant (1985); Chant and McMurtry (1994).

10.3 Life history and biology

The life cycle consists of the egg, larva, protonymph, deutonymph and adult stages. Eggs generally require very high humidity for hatching (e.g. 90% to 100% RH). Larvae of some species never feed, but larvae of some other species must feed for development. Feeding by larvae in some species is facultative. Phytoseiids develop faster than spider mites. Most species complete development within a week and some *Phytoseiulus* species can complete development from eggs to adults within four days.

Phytoseiids are pseudo-arrhenotokous and mating is required for reproduction, although in a few thelytokous species no males have been found. Sex ratio is female-biased with an approximate 3:1 female:male ratio for many species.

Reproductive rates vary depending on predator species and many other factors and some species produce as many as five eggs per day. On average, members of *Phytoseiulus* (2.8), *Iphiseius* (2.4) and *Neoseiulus* (2.3) produce more eggs per day than members of *Galendromus* (1.8), *Euseius* (1.3), *Phytoseius* (0.9) and *Typhlodromus* (0.9). The oviposition period lasts for 20-30 days and fecundity of most species range between 30 and 40 eggs.

Phytoseiids can reproduce throughout the year in tropical and subtropical areas, and also in greenhouses in temperate areas. In the field in temperate areas, they overwinter in protected habitats as fertilized females. Many species have facultative reproductive diapause, which is induced by short day length and low temperature. Some species or strains have better tolerance to low temperatures than others.

With relatively long legs, phytoseiids move fast and can cover short distances by walking on the surfaces of leaves and stems, webbing of spider mites and ground surface. They can respond to kairomones emitted by the prey and can stay within prey-infested areas of the plant or find new nearby infested areas. Long distance dispersal of phytoseiids is passive by air currents. Before taking off, phytoseiids display standing postures, which have more drag force than in the walking posture in the wind boundary layer.

Phytoseiid mites feed on a variety of food and have developed different feeding habits. Four main life-styles have been identified. Type I includes specialist predators (*Phytoseiulus*) that bear long dorsal setae and are adapted to feed on *Tetranychus* species that produce heavy webbing; they rarely attack spider mites other than Tetranychinae. Type II phytoseiids [e.g. *Neoseiulus californicus* (McGregor)] also have relatively long dorsal setae; they feed on Tetranychinae as well as many other small mites and pollen, although they prefer Tetranychinae. Type III phytoseiids (e.g. *Iphiseius degenerans* Berlese) are generalists that feed on various mites, pollen, and insects, but do not do well on *Tetranychus* with strong webbing. Type IV phytoseiids (*Euseius*) are generalist predators of mites and insects but specialist feeders of pollen; their movement is often hindered by strong *Tetranychus* webbing.

References & suggested further reading. Hussey and Huffaker (1976); Helle and Sabelis (1985); Zhang and Croft (1994); Zhang (1995); McMurtry and Croft (1997).

10.4 Species important in greenhouses

Some 20 species of phytoseiids have been made commercially available for pest control and many of these have been applied on greenhouse plants. Several genera have been recorded from, tried or used in greenhouses: *Phytoseiulus, Neoseiulus, Galendromus, Typhlodromus, Typhlodromalus, Amblyseius* and *Euseius*. Seven species have been proven to be of importance in greenhouses and have been commonly used: *Phytoseiulus persimilis, Neoseiulus cucumeris, Neoseiulus barkeri, Neoseiulus californicus, Neoseiulus fallacis, Iphiseius degenerans* and *Galendromus occidentalis*. These genera and species can be separated using the key in Figs 10.1-10.6.

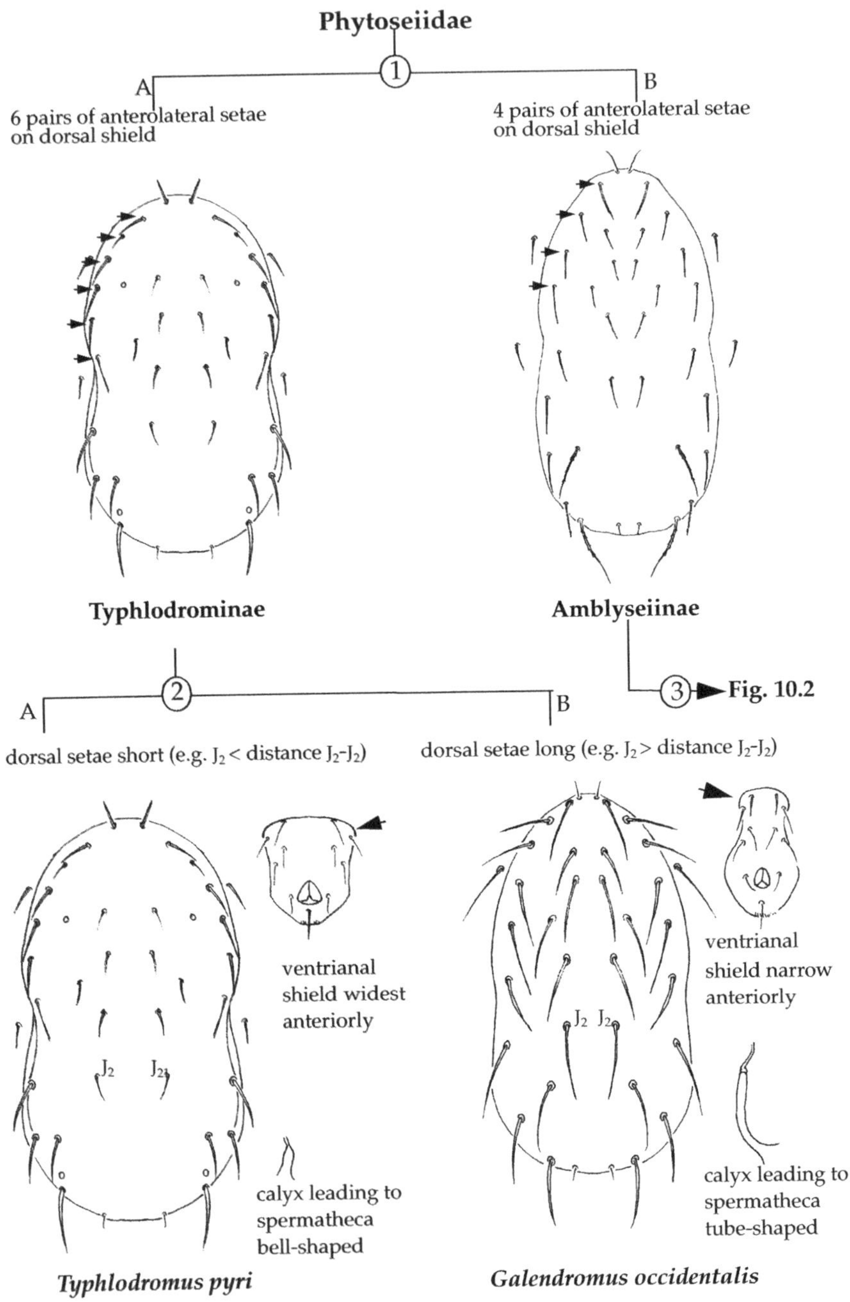

Fig. 10.1 Illustrated key to greenhouse phytoseiid mites. Part I.

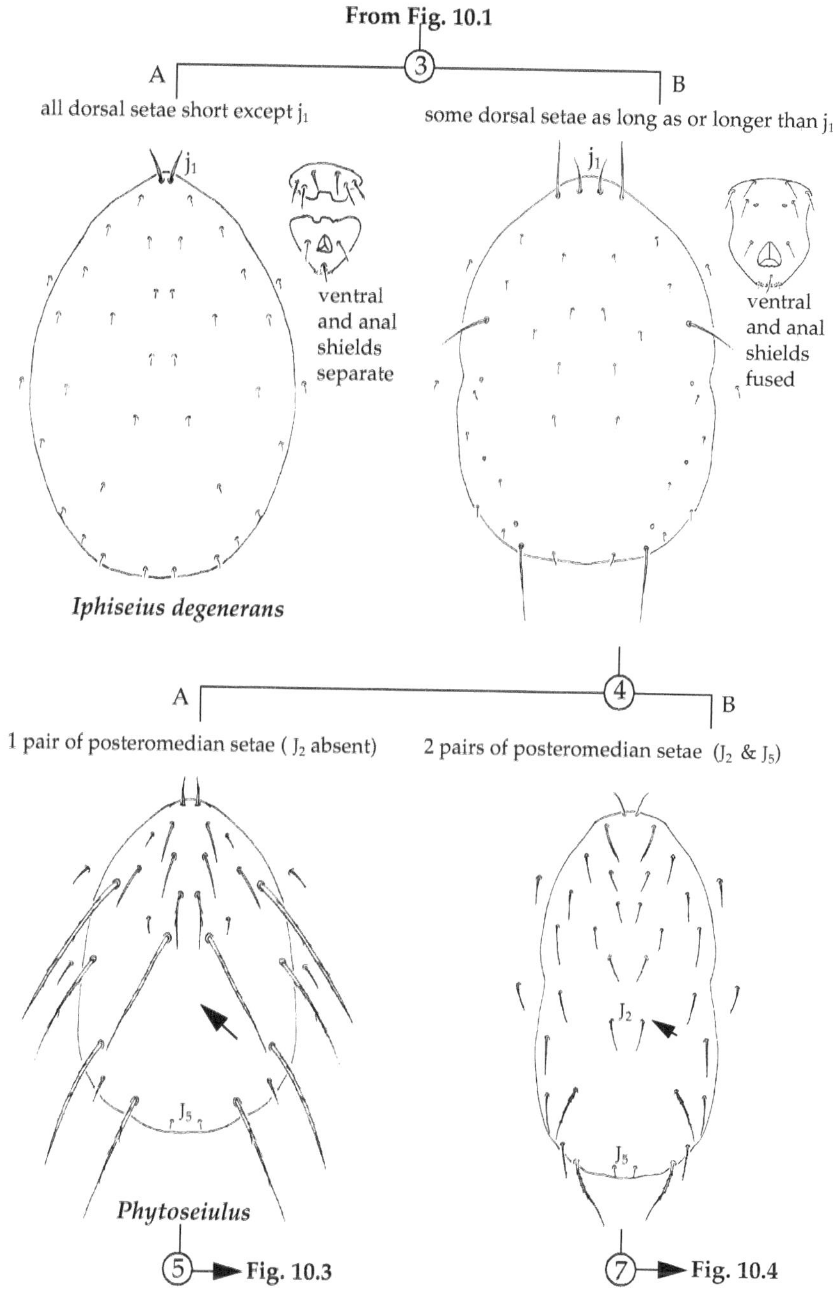

Fig. 10.2 Illustrated key to greenhouse phytoseiid mites. Part II.

From Fig. 10.2

5

A — dorsal shield with 12 pairs of setae; setae j_5 and S_5 absent

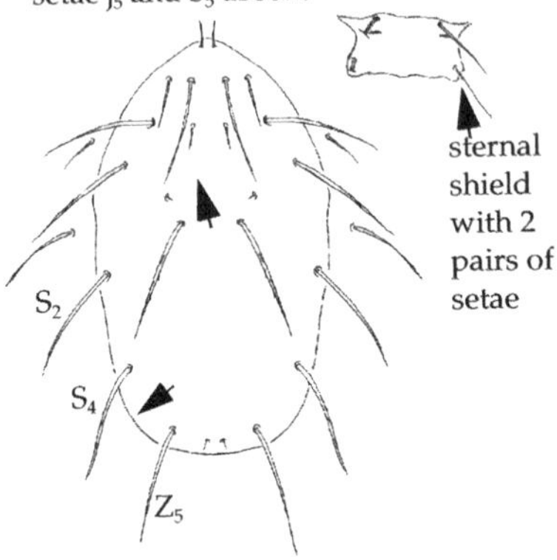

Phytoseiulus longipes

B — dorsal shield with 14 pairs of setae; setae j_5 and S_5 present

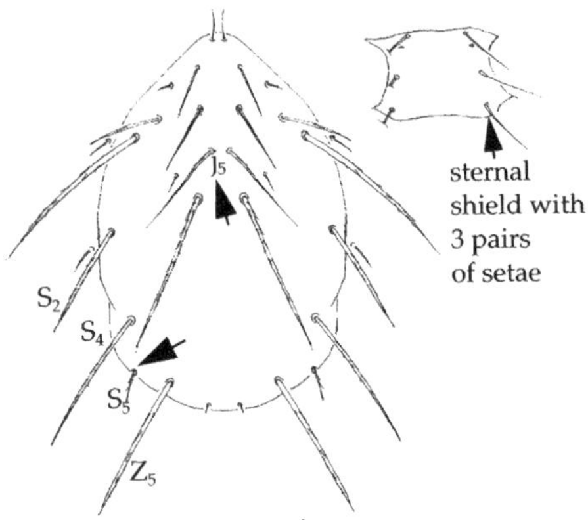

6

A — setae J_6 extending well beyond bases of Z_4

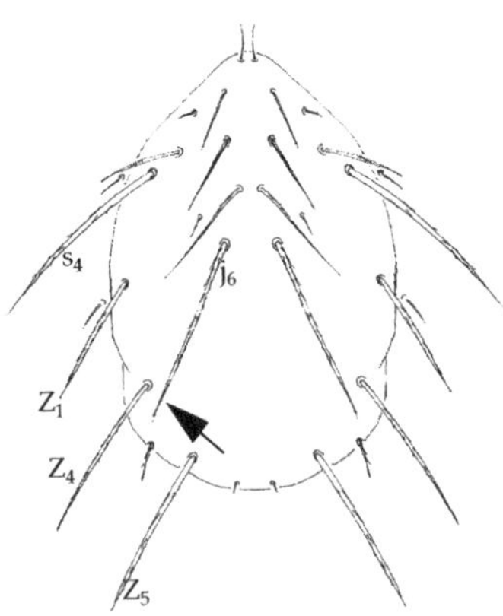

Phytoseiulus persimilis

B — setae J_6 shorter, just reaching bases of Z_4

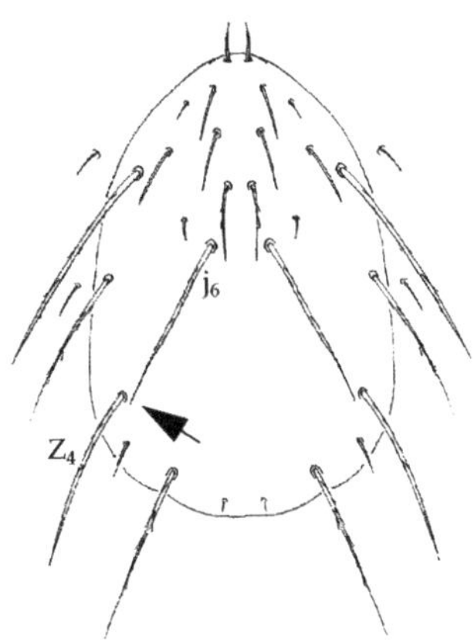

Phytoseiulus macropilis

Fig. 10.3 Illustrated key to greenhouse phytoseiid mites. Part III.

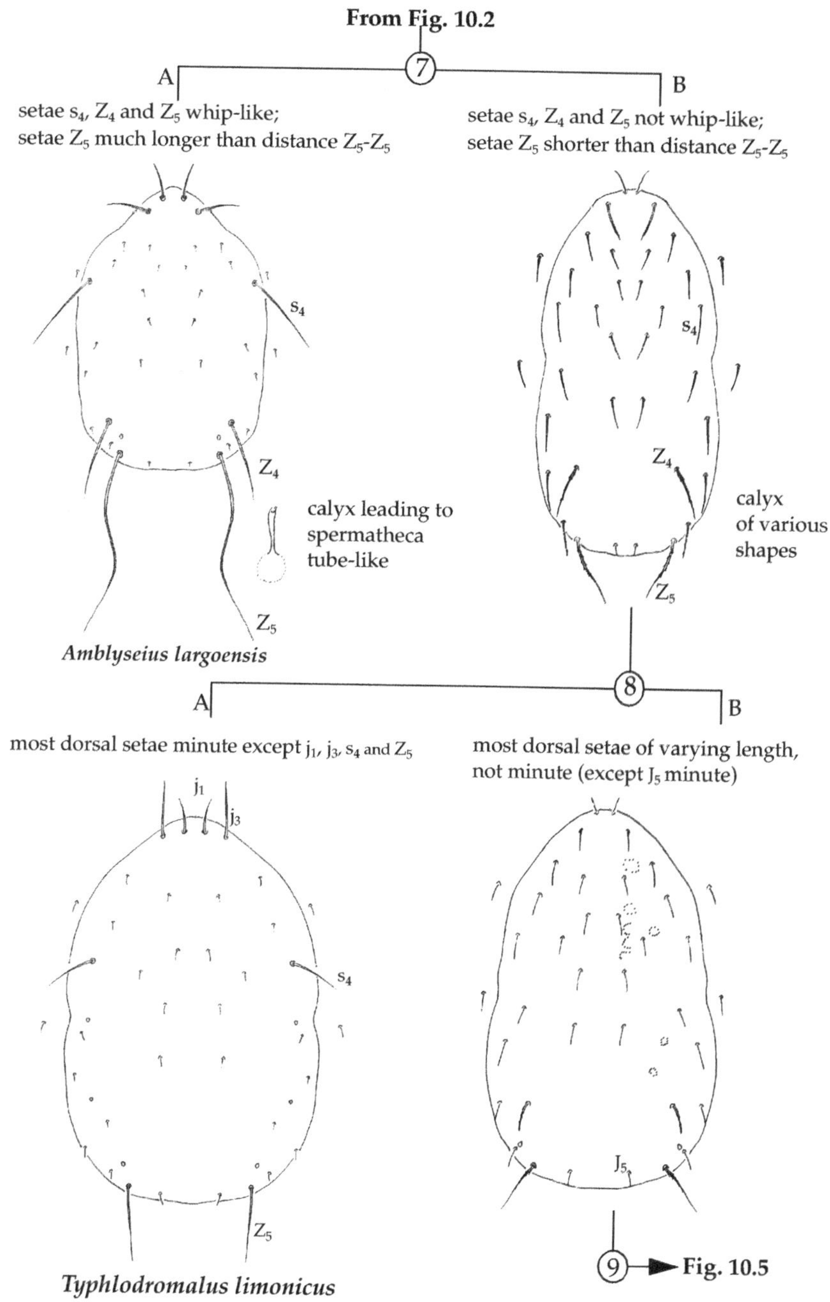

Fig. 10.4 Illustrated key to greenhouse phytoseiid mites. Part IV.

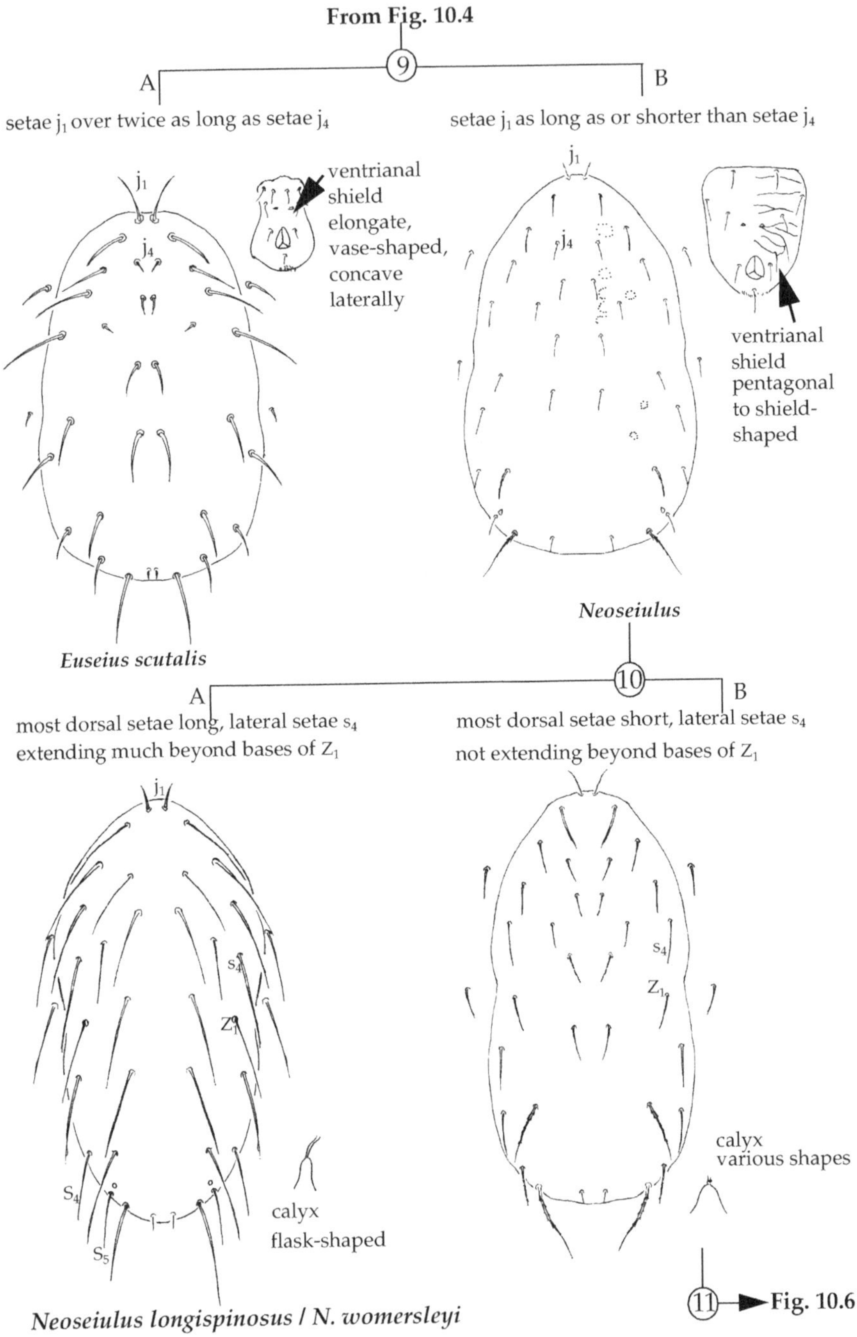

Fig. 10.5 Illustrated key to greenhouse phytoseiid mites. Part V.

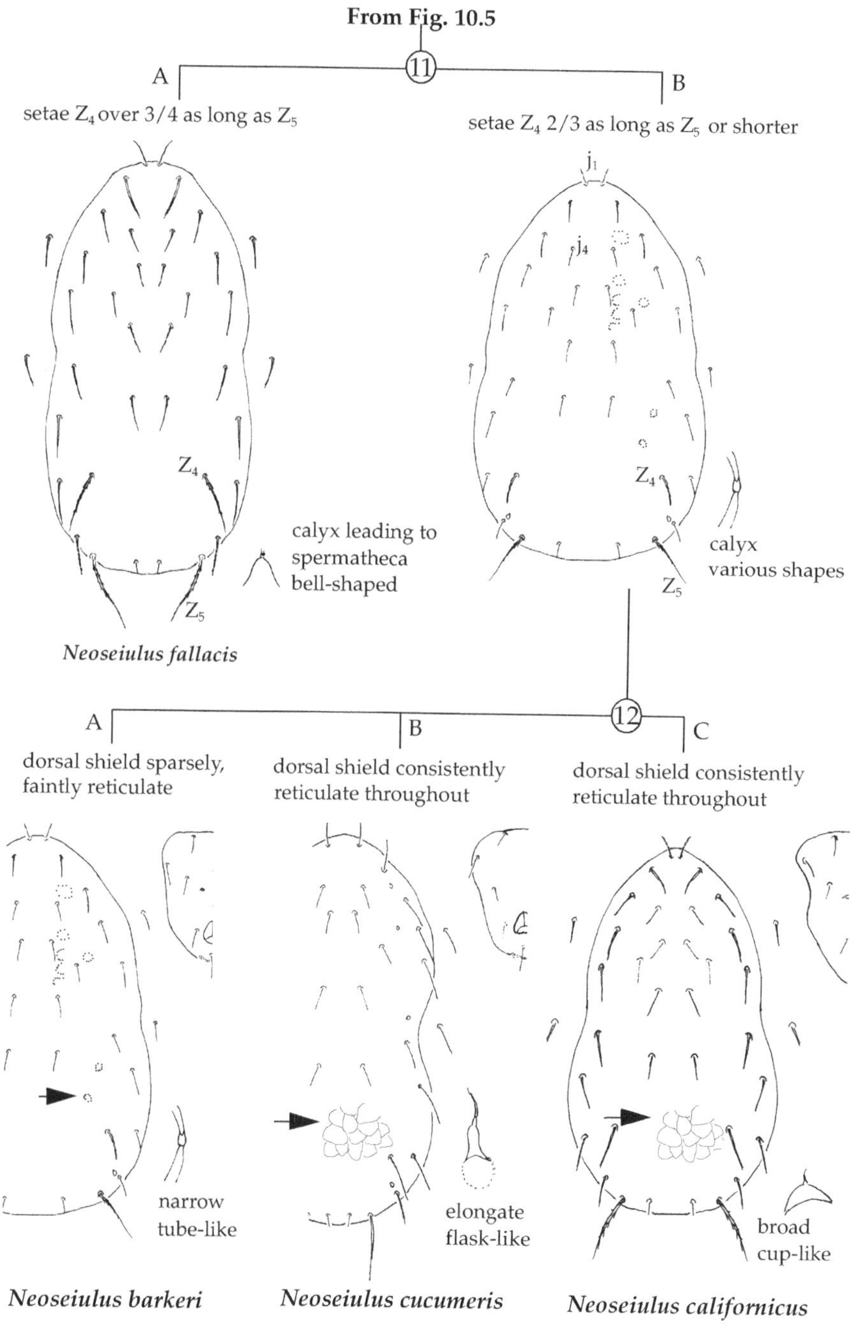

Fig. 10.6 Illustrated key to greenhouse phytoseiid mites. Part VI.

10.4.1 *Phytoseiulus persimilis* Athias-Henriot

10.4.1.1 Common Name

No official common name has been used, but it is known as the Chilean predatory mite because its first use in biological control was in Germany where it was accidentally introduced in orchid roots from Chile.

10.4.1.2 Distribution And Prey

This mite was first discovered on roses grown in greenhouses in Algeria in 1957 and is now known from many places with Mediterranean climates. It has been introduced to many countries and is now used throughout the world for the control of *Tetranychus* spider mites on crops such as cucumber, pepper, tomato, aubergine, strawberry and cut flowers in greenhouses.

This mite is a specific predator of *Tetranychus* spider mites and shows reduced reproduction and survival on other spider mites and phytophagous mites (Type I). It has been collected in association with tetranychine spider mites of the genera *Tetranychus, Eutetranychus* and *Panonychus*, and other predatory mites such as *Neoseiulus californicus* (McGregor) and *Iphiseius degenerans* (Berlese).

References & suggested further reading. Takahashi and Chant (1993a).

10.4.1.3 Appearance And Diagnostic Characters

The eggs are oval and yellowish pale in colour. Larvae and nymphs are pale and translucent at first, but become yellowish to orange after feeding. Adults are ovoid initially, but become pyriform when fully fed and gravid. They are orange to brownish in colour.

Adult females are slightly larger than males. The dorsal shield is about 320 µm long and reticulate laterally, bearing 14 pairs of setae. Setae j_6, Z_1, Z_4, Z_5 and s_4 are >100 µm in length (Fig. 10.3.6A). The sternal, genital and ventrianal shields are strongly reticulate. The ventrianal shield has three setae and lacks preanal setae. The fixed digit of chelicera has seven or eight teeth. Basitarsus IV has a smooth macroseta.

References & suggested further reading. Takahashi and Chant (1993b).

10.4.1.4 Life History And Biology

Development from the egg to adult takes 3.6 days for males and 4.1 days for females at 26°C. An adult female can consume ten to 20 *Tetrancyhus* spider mite eggs per day and lay as many as five eggs per day and up to 80 eggs during her life. The sex ratios of offspring are often highly female-biased (>80% daughters). These life history traits vary with prey species and abundance. When prey density is low, the proportion of daughters produced will approach the even rate of 50%. Host plants also affect the life history of *P. persimilis*, which has a shorter lifespan and lower rates of oviposition on tomato leaves than on bean (*Phaseolus vulgaris*) leaves. Exudates from glandular hairs of tomato are known to be toxic to *P. persimilis*.

The larvae of *P. persimilis* do not feed. The two nymphal stages of a *P. persimilis* attack, on average, 15 eggs of *Tetranychus urticae* and 13 eggs of *T. cinnabarinus*, and consume over 90% of the attacked eggs.

At 70-75% RH, eggs and nymphs of *P. persimilis* can tolerate a temperature of 35 °C for up to 16 hours, but can not survive at 40°C.

In greenhouses where temperatures fluctuate between 11.5 and 23.5°C and relative humidity ranges from 36 to 95%, the egg to adult development is completed in approximately 11 days, while the preoviposition period of the female is 2.8 days.

References & suggested further reading. McClanahan (1968); Takafuji and Chant (1976); Friese and Gilstrap (1985); Guo and Dong (1987); Takahashi and Chant (1992); Gillespie and Quiring (1994); Castagnoli *et al.* (1998); Toyashima and Amano (1998).

10.4.1.5 Use In Biocontrol

Phytoseiulus persimilis can provide effective control of *T. urticae* in greenhouses but control is unsatisfactory at very high temperatures and low humidities. In greenhouses in many countries, *Tetranychus* mites can be controlled by *P. persimilis* using several release systems and the use of pesticides against them is greatly reduced or completely discontinued.

Cucumber

Phytoseiulus persimilis has been used successfully for greenhouse spider mite control on cucumbers in many countries and for many years. Different release methods have been developed depending on whether monitoring of pest population is undertaken and the compatibilities of biological control with environmental conditions and other control measures. Mathematical models have been developed to find optimal management guidelines.

Earlier studies in the 1960s in the UK on spider mite control using *P. persimilis* on this crop led to great success of this species in the greenhouse industries. The control programme uses a leaf-damage index system (a score of 4 for a leaf covered with mites). When the mean leaf-damage index reaches 0.4, two predators are released to every other plant near the infestation site and predators will overrun the spider mites in 30 days, eliminate them in seven weeks and continue to survive for at least three weeks. In the second year during late April and early May, ten to 20 spider mites per plant should be introduced first and two predators introduced to every five plants ten days later. This 'pest in first' programme will ensure successful control of spider mites during the growing season.

This "pest in first" method was later optimized in Europe during the 1980s with the development of a mathematical modeling of the population dynamics of *T. urticae* and *P. persimilis* on cucumber under greenhouse conditions. The optimum distance between the artificial foci is determined based on the dispersal ability of the predators, the optimal initial numbers of spider mites and predatory mites for establishing artificial foci, and the economic threshold of the pest. As a result, the optimization of the "pest in first" method enables the number of *T. urticae* introduced to be reduced 20-fold and that of *P. persimilis* three- to fourfold.

In Austria, *T. urticae* are controlled by *P. persimilis* but the "pest in first" method has not been demonstrated conclusively as advantageous. In Bulgaria, spider mite populations are monitored and *P. persimilis* is introduced onto cucumbers at a predator:prey ratio of 1:20 when there are five to six spider mites per leaf and 20% leaf damage. This method has been successfully used against tetranychid mites. In curative control, timing and rate of predator release is important. Releasing too many predators may result in the lack of food for predators and releasing too few may result in poor control.

Extreme temperature and humidity in greenhouses can affect predator performance. In greenhouses in Egypt, for example, relative humidity can be within 22-68%, and the temperature fluctuates and can get as high as 43°C, which can kill *P. persimilis*. The greenhouse conditions can be modified to favour *P. persimilis*. For example, regular overhead misting can reduce populations of *T. urticae* (which do well under dry and hot conditions) and promote the growth of *P. persimilis* on cucumber greenhouses. A study in Ohio, USA has showed that the Western flower thrips *Frankiniella occidentalis* can also be significantly reduced by a regular misting programme.

The role of *P. persimilis* in IPM has been expanded through the development of resistant strains. In the former USSR, strains resistant to malathion, pirimiphos-methyl, high-temperature, and high temperature plus organophosphorus compounds, respectively, have been developed for use in the control of *T. urticae* on cucumbers in greenhouses. In Japan, a resistant strain of *P. persimilis* can control *T. urticae* on cucumbers under seven sprays of fungicides and four sprays of fenitrothion. The control is successful when they are introduced at the rate of one *P. persimilis* female per ten *T. urticae* females.

In Turkey, where *Tetranychus cinnabarinus* also attacks cucumbers in greenhouses, *P. persimilis* provides effective control when released at the 1:5 predator:prey ratio.

References & suggested further reading. Parr and Hussey (1966); Gould *et al.* (1969); Parr (1969); Hussey and Scope (1985); Pruszynski *et al.* (1985); Lindquist *et al.* (1987); Loginova *et al.* (1987); Markkula *et al.* (1987); Nakao *et al.* (1987); Blumel (1989); Golovkina and Zvereva (1991); Oncuer *et al.* (1994); Cheng *et al.* (1996); Saito *et al.* (1996); Tomczyk *et al.* (1996); Kazak *et al.* (1997); Yoldas *et al.* (1999).

Tomato

Biological control of spider mites on tomato is generally less successful than on cucumber. One reason is the interference of predator searching behaviour by the exudating hairs on tomato leaves and stems.

A practical programme similar to that used on cucumber has been used in the UK. When the leaf-damage index reaches 0.2, five predators should be introduced to every fifth plant. For a new crop, one-fifth of the plants should be inoculated with 30 spider mites three weeks before planting. Four predators should then be introduced on every infested plant ten days later.

In Bulgaria, *P. persimilis* has been successfully used against spider mites, when it is introduced on to tomatoes at the predator:prey ratio of 1:10 when there are two to three spider mites per leaf.

In greenhouses in Belgium, temperatures frequently rise much above 30°C with mean RH below 57% in summer. There is a higher density of glandular trichomes on leaflets at higher temperatures and a larger percentage of the predators is stuck to leaflet trichomes, especially on some cultivars. Increasing the humidity by using a humidifying system and reducing the temperature by whitewashing the roof can reduce mite damage. However, it is better to use predators that are reared on tomato rather than on beans. Strains adapted to tomato have been selected and made available commercially for spider mite control on tomatoes.

Pesticide-resistant strains of *P. persimilis* have also been used on greenhouse tomatoes. In Bulgaria, a strain resistant to organophosphorous pesticides is used in integrated control of tomatoes in the greenhouses. Predators can be used for spider control while chemical control is employed against other pests.

A relatively new approach has been the use of bean (*Phaseolus vulgaris*) as indicator plants for spider mites in greenhouse tomato production.

Although spider mite population growth rates are the same on both host plants, spider mites can became established five weeks earlier on beans because of the lower temperature threshold on beans (7°C lower) than on tomatoes. *T. cinnabarinus* can be controlled when *P. persimilis* is released at the threshold of 12 spider mites per tomato leaf. For some reason, *P. persimilis* is more efficient on the tomatoes than on the beans, which helps to keep the two predator–prey systems on the two hosts out of synchrony, giving better persistence of the system.

References & suggested further reading. Hussey and Scope (1985); Loginova *et al.* (1987); Nihoul (1993a,b, 1994); Nihoul and Hance (1993); Oncuer *et al.* (1994); Atanasov (1995); Berlinger *et al.* (1996); Drukker *et al.* (1997); Schelt and Altena (1997); Yoldas *et al.* (1999).

Capsicum

When *P. persimilis* is introduced to sweet peppers at the predator:prey ratio of 1:10 when the initial spider mite density is two to three per leaf, successful control can be achieved. In Bulgaria, *P. persimilis* provides effective control of *T. urticae* and *T. turkestani* on pepper when released at a rate of 70,000-80,000 individuals per hetcare, reducing the need for chemical control. In Poland, *P. persimilis* is effective against both *T. urticae* and *T. cinnabarinus* developing separately or together on plants of two cultivars of greenhouse sweet pepper. In unheated greenhouses in Sicily, releases of *P. persimilis* from February onwards provides good control of *T. urticae* on chilli, but control can be disrupted by the application of chinomethionat against powdery mildew from March.

References & suggested further reading. Atanasov *et al.* (1983); Pruszynski *et al.* (1985); Loginova *et al.* (1987); Kropezynska and Tomczyk (1996).

Other food crops

In greenhouse strawberries in Italy, *P. persimilis* gives effective control of *T. urticae* when released at the predator:prey ratio of about 1:10 at the initial spider mite density of one to two per leaf. In aubergine crops grown in plastic greenhouses in Tunisia, one release of *P. persimilis* at the rate of one predator for ten *T. urticae* can completely eliminate prey in seven weeks. In Bologna, Italy, *P. persismilis* is also used to control *T. urticae* on aubergine in greenhouses. In greenhouse grown melons in Spain, *P. persimilis* should be released at the rate of $10/m^2$ at the foci of infestation when *T. urticae* first appears. In Japan, *P. persimilis* is effective against *Tetranychus kanzawai* in greenhouse-grown grapes (see 4.4.3.2).

References & suggested further reading. Battaglia *et al* (1990); Bonomo *et al.* (1991); Chermiti (1992); Baraja *et al.* (1996); Saito *et al.* (1996); Castaldi (1999).

Ornamental plants

Ornamental plants are grown for cosmetic purposes and can tolerate lower levels of mites than other plants. *P. persimilis* has been used to control spider mites on a variety of ornamental plant species with success, and is sometimes used with spot treatment of mite infestations using acaricides.

On greenhouse roses, *T. urticae* has been effectively controlled by *P. persismilis* in many countries. When *P. persimilis* reachs the level of one predator for every ten spider mites, it can eliminate spider mites in a few weeks time and then disperse themselves. In very large greenhouses, *P. persimilis* can persist in the lower canopies and can tolerate spot treatment of upper foliage using selective chemcials such as abamectin.

Gerbera, being more bushy with a more humid microenviroment, favours the development of *P. persimilis*. In greenhouses in Sicily, Italy, natural populations of *P. persimilis* move in and provide natural control of *T. urticae* on *Gerbera*, provided pesticides toxic to the predator are not used.

In the UK, *T. urticae* on chrysanthemums is successfully controlled by releasing ten *P. persimilis* females per plant three to four weeks after treatment of the plants with aldicarb.

In China, *T. urticae* on *Salvia splendens, Ageratum conyzoides, Zantedeschia aethiopica* and *Pelargonium lateripes* in greenhouses are successfully controlled by releases of *P. persimilis* either at the seedling stage or when mites just begin to increase. Such timing of releases is both effective and economic. The release rate ranges from one to 50 mites/plant, depending on the pest density and plant species and size. A second release is sometimes necessary when the spider mite populations begin to increase on some plants.

In Iowa, USA, *P. persimilis*, along with *Neoseiulus californicus*, is released to control *T. urticae* on greenhouse-grown poplar (*Populus* spp.) with spot treatment using pestcides when needed. Spider mites are suppressed to acceptable levels and pest management costs are reduced by 81% compared to chemical control.

In Poland, *P. persimilis* provides good control of *T. urticae* on orchids in greenhouses.

In New Zealand, *P. persimilis* provides good control of *T. urticae* on *Cymbidium*, but is not effective against *T. urticae* on greenhouse carnations due to its inability to maintain traction on the waxy surface of the leaves and stems.

References & suggested further reading. Simmonds (1972); Vacante and Firullo (1983); Hussey and Scope (1985); Pruszynski *et al.* (1985); Dong *et al.* (1986); Vacante and Garzia (1987); Blumel (1990); Beck *et al.* (1993); Smith *et al.* (1993); Zhang and Sanderson (1995); Workman and Martin (2000).

10.4.2 *Neoseiulus cucumeris* (Oudemans)

10.4.2.1 Common Name

No official common name has been used, but it is also known as *Amblyseius cucumeris*.

10.4.2.2 Distribution And Prey

This species is known throughout the world. This species is a generalist predator (Type III), feeding on pollen, small insects and small mites.

References & suggested further reading. Moraes *et al.* (1986).

10.4.2.3 Appearance And Diagnostic Characters

The eggs are oval and translucent. Larvae and nymphs are pale to yellowish. The adult females are larger than the males, measuring about 400 µm. The dorsal shield is reticulate throughout and bears 17 pairs of setae (Fig. 10.6.12B). Most setae are shorter than the distance between setal bases in the same series. Setae Z_5 are slightly serrated. The calyx of the spermathecal apparatus is elongate flask-like. The ventrianal shield is quadrate, broad anteriorly. Leg IV has three microsetae.

References & suggested further reading. Collyer (1982); Beard (2001).

10.4.2.4 Life History And Biology

At 25°C, the egg to adult development is completed in eight to nine days when feeding on thrips larvae and about seven days when feeding on acarid mites. Females produce an average of 53 eggs during the oviposition period at a rate of 1.9 eggs/day. The intrinsic rate of increase is 0.203 females/female per day and the population is able to double in 3.41 days.

This species can feed on pollen and the availability of pollen on greenhouses enhances the development and reproduction of *N. cucumeris*, although pollen feeding reduces the predation rate on prey.

In temperate areas, the effectiveness of *N. cucumeris* in autumn and winter may be limited by the reproductive diapause induced under short-day conditions. The critical day-length for inducing diapause at 22°C is 12.45 hours and most *N. cucumeris* undergo diapause only when exposed to diapause-inducing conditions throughout their juvenile development. A New Zealand strain with low incidence of diapause has been selected and improved, and is now widely used in greenhouses around the world.

References & suggested further reading. Gillespie and Ramey (1988); Castagnoli (1989); Castagnoli *et al.* (1990); van Rijn and Sabelis (1990, 1993); Morewood and Gilkeson (1991); Cloutier *et al.* (1995); Houten *et al.* (1995).

10.4.2.5 Use In Biocontrol

This mite is an effective predator of some tarsonemid mites (see Chapter 7) and some spider mites that do not produce webbing (see Chapter 4). It has also been tested against *Bemisia tabaci* in the laboratory. It can complete its development to the adult stage feeding on a combination of eggs and first- and second-instar larvae of *B. tabaci* with a 72% survival rate and can also reproduce on eggs and larvae of *B. tabaci*. An adult *N. cucumeris* can consume an average of 6.6 eggs, 1.9 first-instar larvae or 0.9 second-instar larvae of *B. tabaci* per day. This species should be tried for whitefly control in greenhouses.

N. cucumeris has been used most successfully against thrips on vegetables and ornamentals in greenhouses. Often repeated releases and high predator:prey ratios are required for adequate control. Preventive releases are also very effective. Its performances on different plants are varied.

References & suggested further reading. Nawar and El-Sherif (1993).

Cucumber

In the former USSR, larvae of *Thrips tabaci* are controlled on cucumbers in greenhouses when *N. cucumeris* is released at predator:prey ratios of 1:2. In Turkey, effective control is achieved by releasing predators at the rate of four to five individuals per plant as soon as thrips are detected on host plants. In the UK, good establishment result is possible by either a single release of 250 predators/plant or three consecutive releases of 50 predators/plant at two-weekly intervals, starting at the first sign of infestation. *N. cucumeris* can reduce populations of *T. tabaci*, but can not eliminate the population of the thrips. In Canada, *N. cucumeris* provides effective control of *T. tabaci* and *Frankliniella occidentalis* on seedless cucumber in greenhouses in British Columbia. Adult predators can persist on plants for seven weeks in the virtual absence of thrips prey, and can increase numerically in response to increases in thrips populations.

Neoseiulus cucumeris can keep *Frankliniella occidentalis* populations at a low level, but sometimes do not provide effective control, despite repeated releases. This is overcome by the development of controlled release system (CRS), which provides better distribution of the predator than the traditional loose bran system. The CRS consists of a specially formulated bran-based population of *N. cucumeris* in a waxed paper pack of specific porosity, and gives more rapid establishment. The predators continue to breed and emerge from the pack for at least six weeks. The technique provides better control of *F. occidentalis* and is also less expensive than the traditional method. The best control is prevention. When *N. cucumeris* is introduced in good quality culture sachets immediately after planting, thrips populations do not develop.

References & suggested further reading. Beglyarov and Suchalkin (1983); Bennison (1988); Gillespie (1989); Ramakers *et al.* (1989); Bennison and Jacobson (1991); Higgins (1992); Rodriguez-Reina *et al.* (1994); Grasselly *et al.* (1995); Kazak *et al.* (1999); Jacobson (1995).

Pepper

The effectiveness of *N. cucumeris* in greenhouses is inconsistent. In The Netherlands, control of thrips on peppers is less successful than on cucumber. In the USA, two introductions of ten to 25 predators per plant during the growing season are not effective against *F. occidentalis* in greenhouses. In Spain, inundative releases of 50-400 predators/plant on different occasions and controlled releases are unable to control *F. occidentalis* on sweet pepper in unheated greenhouses. However, the combined use of *N. cucumeris* and predatory bugs (*Orius*) can provide effective control of *F. occidentalis* on sweet peppers in greenhouses.

The effectiveness of *N. cucumeris* against thrips of the genus *Thrips* are more consistent. In greenhouse sweet peppers in Japan, *N. cucumeris* released three times at one-week intervals starting three days after planting reduces *T. palmi* population to one-third to one-fifth of its original size for six weeks after planting. In New Zealand, three releases of 10-140 *N. cucumeris* per plant against *Thrips tabaci* and *T. obscuratus* result in low populations of *Thrips* and high predator numbers for 12 weeks, with no apparent thrips damage to plants.

References & suggested further reading. Ramakers (1988); Ramakers *et al.* (1989); Higgins (1992); Vacante and Garzia (1993); Workman *et al.* (1994); Dissevelt *et al.* (1995); Kurogi *et al.* (1997); Sanchez *et al.* (1997).

Ornamental plants

Control of *F. occidentalis* on chrysanthemums using *N. cucumeris* has been effective in both Europe and North America. In the USA, three predatory mites per leaf are released when the density of adult thrips is high, and pollen is released with the mites to encourage predator survival when thrips are low in number. In the UK, releasing 100 *N. cucumeris* in bran per m^2 every other week gives effective biological control of *F. occidentalis* on chrysanthemums. In bed-grown year-round chrysanthemums in greenhouses, effective control can be achieved by introducing three predators per plant in the third, fourth and fifth weeks after planting.

In the UK, *N. cucumeris* is also effective against *F. occidentalis* on cyclamen in greenhouses. In the USA, it provides control of *Frankliniella tritici* and *F. occidentalis* on ornamental bedding plants in greenhouses when breeding sachets containing 50 mites are introduced at the rate of 125 sachets per 200 m^2 of growing area.

References & suggested further reading. Hessein and Parrella (1990); Wardlow *et al.* (1991); Buxton and Finlay (1993); Courcy Williams (1993); Stanton (1994).

10.4.3 Other phytoseiid species

10.4.3.1 *Neoseiulus barkeri* Hughes

Also known as *Amblyseius barkeri*, this is a widespread and polyphagous species. It feeds on pollen, many small mites and also small insects such as thrips and whitefly immatures.

The eggs are oval and translucent, about 90 µm long. Immature stages are pale to yellowish, but adults are darker in colour, often pale-brown. Adult females are about 400 µm long. The dorsal shield bears 17 pairs of setae (Fig. 10.6.12A). There are some faint reticulate markings on the dorsal shield. The ventrianal shield is subquadrate. Leg IV has one macroseta on the basitarsus.

Development of *N. barkeri* occur between 15-35°C and a relative humidity of >90%. With *Tyrophagus putrescentiae* as prey, the egg to adult development takes six days. Adult lifespan increases with rising relative humidity. Optimal conditions for *A. barkeri* are 25-30 °C and 90% RH or higher. Females of *A. barkeri* fed with *T. urticae* lay more eggs than those fed with *T. putrescentiae*. When *A. barkeri* is fed on *T. urticae*, the motile juvenile stages are usually eaten first. Females generally consume more than males. Females lay a maximum of 1.96 eggs per day. Adult life span averages 161 days at 26°C.

Like *N. cucumeris*, *N. barkeri* goes into reproductive diapause under short-day conditions. A strain from The Netherlands with low incidence of diapause has been selected and improved, without changes in life history traits. *A. barkeri* has lower rates of predation and oviposition than *N. cucumeris* when feeding on *F. occidentalis*.

This species can provide effective control of the broad mite on pepper (Chapter 6). Like *A. cucumeris*, it has been available commercially for thrips control, but its effectiveness is less consistent than *N. cucumeris* in inundative biological control. This species feeds and reproduces on immature *Bemisia tabaci* and has been suggested as a biocontrol agent of the whitefly.

References & suggested further reading. Hughes (1976); Ramakers (1988); Baier and Karg (1992); Fouly and El-Laithy (1992); Rodriguez-Reina *et al.* (1994); Houten *et al.* (1995); Jarosik and Pliva (1995); Momen (1995, 1996); Nomikou *et al.* (2001).

10.4.3.2 *Neoseiulus californicus* (McGregor)

Also known as *Amblyseius californicus*, this species was originally described from California. It is also known from Central and South America and southern Europe, and has been introduced to Europe, Asia and Africa for biological control. This is a Type II phytoseiid species, mainly feeding on *Tetranychus* spider mites and also other mites and pollen. Like Type III phytoseiids, it can also feed on small insects.

The eggs are oval and translucent. The adults are yellow. Adult females are about 350 µm long. The dorsal shield is covered with a reticulate pattern and bears 17 pairs of dorsal setae. Setae Z_4 are armed with short barbs and are longer than other setae. The calyx of the spermathecal apparatus is characteristically broad cup-shaped (Fig. 10.6.12C). The ventrianal shield is wasted laterally.

The egg-to-egg generation time is 9.5 days at 25 ± 1°C when reared on broad mites, two days longer than when reared on *Tetranychus urticae* under the same conditions. The proportion of females in the offspring is 51.2%, and each female lays an average of over two eggs per day. The larvae are active and feed on prey. The nymphs of *N. californicus* can attack 13 eggs of *T. urticae* and 11 eggs of *T. cinnabarinus*, consuming over 86% of the eggs killed. Adult females can resist starvation for over ten days and can resume oviposition when food is available. Female fecundity is 65 eggs when feeding on nymphs of *T. urticae*.

This species is from arid and semi-arid climates and more tolerant to low humidity and low prey density. In perennial greenhouse-grown crops *P. persimilis* and *N. californicus* have complementary effects and a combination of the two can enhance long-term biological control of spider mites.

This species is also an effective biocontrol agent of broad mites and cyclamen mites (see Chapter 6).

References & suggested further reading. Friese and Gilstrap (1985); Moraes *et al.* (1986); Ehara and Amano (1993); Castagnoli and Falchini (1993); Croft *et al.* (1998); El-laithy and Elsawi (1998); Palevsky *et al.* (1999); Schausberger and Walzer (2001).

10.4.3.3 *Neoseiulus fallacis* (Garman)

This species is common in humid parts of the eastern and central USA but is distributed in all major continents of the world, some by artificial introduction for biological control. It is common in orchards. This is a Type II phytoseiid species, mainly feeding on spider mites and also other mites and pollen.

The eggs are oval and translucent. The adults are yellowish. The dorsal shield is covered with a reticulate pattern and bears 17 pairs of dorsal setae. Setae Z_4 and Z_5 are armed with short barbs and are longer than other setae. The calyx of the spermathecal apparatus is bell-shaped (Fig. 10.6.11A). The ventrianal shield is broad anteriorly. Leg IV has three macrosetae.

This species is an effective predator of the cyclamen mite (see Chapter 6) and tomato russet mite (see Chapter 7). Its most significant role in biological control is against spider mites in orchards, although it can also be applied to control spider mites such as *Tetranychus* and *Panonychus* occurring in greenhouses. It is similar to *N. californicus* in biology, but prefers humid habitats and is resistant to many pesticides.

References & suggested further reading. McClanahan (1968); Collyer (1982); Moraes *et al.* (1986); Croft *et al.* (1998); Kwon *et al.* (1998); Beard (2001).

10.4.3.4 *Neoseiulus longispinosus* (Evans) and *Neoseiulus womersleyi* Schicha

The separate identities of these two similar species were only confirmed recently. They are known from Russia, China and Japan, through Southeast Asia to Australasia and Hawaii. Some previous records of *N. longispinosus* may actually refer to *N. womersleyi*. Both species are Type II predators, feeding on many species of mites and also on pollen.

The appearances of the two species are similar. The eggs are oval and translucent. Immatures are pale and adults are yellow. Adult females average about 350 μm. The dorsal shield has 17 pairs of dorsal setae (Fig. 10.5.10A). Except j_1 and S_5, all dorsal setae are long and barbed, extending beyond bases of its next setae. The ventrianal shield is quadrate and bears three pairs of pre-anal setae and two pairs of pores. There is one pair of metapodal plates. The calyx of the spermathecal apparatus is flask-shaped. Setae S_5 are barbed and as long as setae S_4 in *N. womersleyi*, but smooth and much shorter than S_4 in *N. longispinosus*.

Immature development is completed in five days at 28°C, with *N. womersleyi* being slightly faster than *N. longispinosus*. During the first ten days of adulthood, *N. wormersleyi* adult females consume 32.07 *T. kanzawai* eggs per day and produce 3.07 eggs per day, whereas *N. longispinosus* females consume 26.43 eggs per day and lay 2.92 eggs per day.

Both species are common on low-growing outdoor plants and have rarely been used in greenhouses as frequently as *N. californicus*. *N. womersleyi* has been found naturally in greenhouse grapes in Japan but is unable to provide effective control. Both species have potential to be used along with *P. persimilis* for spider mite control. *N. longispinosus* can feed on the broad mite. Females lay an average of 27.5 eggs during an oviposition period of 16.1 days, consuming 11.7 larvae, 9.3 nymphs or 5.1 adults of broad mites per day during this time. Predator larvae consume 3.8, 1.4 and 0 prey larvae, nymphs and adults, respectively, while protonymphs consume 9.2 larvae, 7.9 nymphs and 3.2 adults.

References & suggested further reading. Collyer (1982); Moraes *et al.* (1986); Hariyappa and Kulkarni (1988); Ashihara *et al.* (1992); Ho *et al.* (1995); Kwon *et al.* (1998); Toyoshima and Amano (1998); Beard (2001).

10.4.3.5 *Galendromus occidentalis* (Nesbitt)

Also known as *Metaseiulus occidentalis* and *Typhlodromus occidentalis,* this species is an important species for the control of spider mites in orchards in North America. It has been recorded from Europe, Asia and Australasia, mostly due to introductions. It is a Type II predator, preying mainly on tetranychine spider mites and also on other small mites.

The eggs are oval and translucent. Immatures are whitish to pale. The adults are pale to yellow. Adult females have a reticulated short shield, measuring about 330 µm and bearing 17 pairs of setae, with six pairs anterior lateral setae (Fig. 10.1.2B). The peritreme is very short. The ventrianal shield is longer than wide, narrow anteriorly, slightly reticulate and bearing four pairs of pre-anal setae. The calyx of the spermathecal apparatus is elongate tube-like (Fig. 10.1.2B).

Development occurs above 10.7°C. Development from the egg to adult is completed in one week at 25°C. Immature *G. occidentalis* kills an average of 10.4 eggs of *T. urticae* and 12.6 eggs of *T. cinnabarinus*, consuming >79% of the eggs killed and requiring about four days to complete development after emergence. Adult females may lay an average of 2.88 eggs per day and 43.8 eggs during a life time. The proportion of female offspring is about 0.677.

This species is an effective predator of the twospotted spider and other spider mites in orchards and has developed resistance to many pesticides. In greenhouses, it has been used successfully to control *T. urticae* on roses in the USA and on small apple trees in Australia.

References & suggested further reading. Collyer (1982); Friese and Gilstrap (1985); Field and Hoy (1986); Moraes *et al.* (1986); Bruce-Oliver and Hoy (1990); van de Klashorst *et al.* (1992); Croft *et al.* (1998); Kwon *et al.* (1998).

10.4.3.6 *Iphiseius degenerans* (Berlese)

Also known as *Amblyseius degenerans*, this widespread species is a generalist predator (Type III). It feeds on small mites, insects and pollen.

The adult females of this species are darkish in colour and sub-globular in shape. There are 17 pairs of setae on the dorsal shield. All the dorsal setae are minute except seta j_1. The interscutal membrane near the dorsal shield is sclerotized. The ventral shield and anal shield are separate, unlike the fused ventrianal shield in other species. Leg IV has three macrosetae.

Development from egg to adult takes 5.7 days at 25°C. Females lay 2.2 eggs per day or a total of 68 eggs during 30 days of the oviposition period. The eggs are relatively tolerant to desiccation. Deutonymphs of *I. degenerans* consume 21.1 *Tetranychus pacificus* per day, whereas those of *P. persimilis* consume 7.5 prey per day. It is a fast-moving species.

This species has recently been employed for the biological control of thrips in greenhouses. Releases of ten predators per plant eliminate citrus thrips *Scirtothrips citri* (Moulton) from potted citrus trees in greenhouses in the USA. In The Netherlands, *I. degenerans* is introduced to greenhouse sweet pepper in potted flowering castor bean bushes, which serves as a reservoir for predatory mites to colonize sweet peppers.

References & suggested further reading. Takafuji and Chant (1976); Moraes *et al.* (1986); van Houten and van Stratum (1993); Ramakers and Voet (1996); Grafton-Cardwel *et al.* (1999).

10.4.3.7 Other minor species

Typhlodromus pyri Scheuten is mainly found in orchards in Europe and North America. It is also known in Egypt and was introduced accidentally to New Zealand. It is a generalist predatory mite (Type III). This species is an important predator of *Panonychus ulmi* and also feeds on other spider mites and eriophyid mites. It has been used to control *T. urticae* on greenhouse strawberries in Europe. The predators are released at predator:prey ratio of approximately 1:10 when *T. urticae* is 9.7 active stages/trifoliate leaf in the beginning of April. The predator gives good control up to and following fruiting in mid-May.

Typhlodromalus limonicus (Garman & McGregor) is widespread in the New World. It is also known in New Zealand and was introduced to Africa. It is a Type III phytoseiid and has recently been explored for control of thrips in greenhouses. On cucumbers in greenhouses, it performs better than *N. cucumeris*. It is yet to be made available commercially for biological control.

Euseius scutalis (Athias-Henriot) is a North African and Middle-Eastern species, feeding on a variety of prey (Type IV). It has recently been explored for control of tobacco whitefly, *Bemesia tabaci* (Gennadius) in European greenhouses. When predators are released with a supply of pollen on cucumber plants two weeks before the release of *B. tabaci*, they can significantly reduce whitefly populations, by 16-fold to 21-fold after nine weeks.

Phytoseiulus macropilis (Banks) and *P. longipes* Evans are Type I predators like *P. persimilis*. Both species are similar to *P. persimilis* in biology. Both are available commercially for biological control of *Tetrancyhus* species for outdoor and greenhouse crops. Both species have been shown to be effective against spider mites in greenhouses but are much underused compared to *P. persimilis*.

Amblyseius largoensis (Muma) is a generalist predator (Type III) of worldwide distribution. It is generally an outdoor species. It was recently found in greenhouses in New Zealand, but its effects on spider mites and thrips are unknown.

References & suggested further reading. Hamlen and Poole (1980); Moraes *et al.* (1986); Takahashi and Chant (1992); El-Laithy (1996); van Houten (1996); Zacharda and Hluchy (1996); Nomikou *et al.* (2001).

10.5 References cited

Ashihara, W., Inoue, K., Osakabe, M. and Hamamura, T. (1992) Effectiveness of *Phytoseiulus persimilis* Athias-Henriot (Acarina: Phytoseiidae) as a control agent for the Kanzawa spider mite, *Tetranychus kanzawai* Kishida (Acarina: Tetranychidae) and occurrence of native natural enemies of the spider mite on grapevine in glasshouse. *Bulletin of the Fruit Tree Research Station* 22, 109-130.

Atanasov, N.D. (1995) Integrated control of mites on tomatoes. *Acta Horticulturae* 412, 546-550.

Atanasov, N., Mitkov, A. and Slavov, D. (1983) Biological control of spider mites on pepper in the greenhouse. *Rastitelna Zashchita* 31, 29-31.

Baier, B. and Karg, W. (1992) Investigations into the biology, ecology and efficiency of oligophagous predatory mites with particular reference to *Amblyseius barkeri* (Hughes) (Acarina: Phytoseiidae). *Mitteilungen aus der Biologischen Bundesanstalt fur Land- und Forstwirtschaft Berlin-Dahlem* 281, 1-88.

Baraja, M.J., Gonzalez, S. and Montalban, C. (1996) Integrated management in a staked glasshouse melon crop. *Horticultura, Revista de Hortalizas, Flores y Plantas Ornamentales* 113, 29-32.

Battaglia, D., Borriello, M. and Spicciarelli, R. (1990) Biological control of *Tetranychus urticae* Koch by *Phytoseiulus persimilis* Athias-Henriot on protected strawberry in the Metapontum area. *Informatore Fitopatologico* 40 (7-8), 44-46.

Beard, J.J. (2001) A review of Australian *Neoseiulus* Hughes and *Typhlodromips* de Leon (Acari: Phytoseiidae : Amblyseiinae). *Invertebrate Taxonomy* 15, 73-158.

Beck, N.G., Workman, P. and Martin, N. (1993) IPM for *Cymbidium* orchids in New Zealand. *Bulletin OILB/SROP* 16, 12-15.

Beglyarov, G.A. and Suchalkin, F.A. (1983) A predacious mite - a potential natural enemy of the tobacco thrips. *Zashchita Rasteni* 9, 24-25.

Bennison, J.A. (1988) Integrated control of thrips on cucumber in the United Kingdom. *Mededelingen van de Faculteit Landbouwwetenschappen, Rijksuniversiteit Gent* 53(3a), 961-966.

Bennison, J.A. and Jacobson, R. (1991) Integrated control of *Frankliniella occidentalis* (Pergande) in UK cucumber crops - evaluation of a controlled release system of introducing *Amblyseius cucumeris*. *Mededelingen van de Faculteit Landbouwwetenschappen, Rijksuniversiteit Gent* 56(2a), 251-258.

Berlinger, M.J., Dijk, B.L., Dahan, R., Lebiush-Mordechi, S. and Taylor, R.A.J. (1996) Indicator plants for monitoring pest population growth. *Annals of the Entomological Society of America* 89, 611-622.

Blümel, S. (1989) The use of *Encarsia formosa* (Gah.) and *Phytoseiulus persimilis* (A.H.) for the control of *Trialeurodes vaporariorum* Westw. (Aleyrodidae) and *Tetranychus urticae* (Koch) on tomato and cucumber under glass. *Pflanzenschutzberichte* 50, 9-18.

Blümel, S. (1990) Use of beneficials for integrated pest control in bouvardias and roses in glasshouses. *Pflanzenschutzberichte* 51, 25-35.

Bonomo, G., Catalano, G., Maltese, V. and Sparta, S. (1991) Biological and integrated control experiments in Marsalese strawberry crops. *Informatore Agrario* 47, 97-100.

Bruce-Oliver, S.J. and Hoy, M.A. (1990) Effect of prey stage on life-table attributes of a genetically manipulated strain of *Metaseiulus occidentalis* (Acari: Phytoseiidae). *Experimental & Applied Acarology* 9, 201-207.

Buxton, J.H. and Finlay, R. (1993) Integrated pest management in AYR chrysanthemums. *Bulletin OILB/SROP* 16(8), 33-41.

Castagnoli, M. (1989) Biology and prospects for mass rearing of *Amblyseius cucumeris* (Oud.) (Acarina: Phytoseiidae) using *Dermatophagoides farinae* Hughes (Acarina: Pyroglyphidae) as prey. *Redia* 72, 389-402.

Castagnoli, M., Del Bene, G., Gargani, E. and Simoni, S. (1990) Possibilita di controllo di Thrips tabaci Lind. e *Frankliniella occidentalis* (Pergande) (Thys. Thripidae) con *Amblyseius cucumeris* (Oud.). *Redia* 73, 53-61.

Castagnoli, M. and Falchini, L. (1993) Suitability of *Polyphagotarsonemus latus* (Banks) (Acari, Tarsonemidae) as prey for *Amblyseius californicus* (McGregor) (Acari, Phytoseiidae). *Redia* 76, 273-279.

Castagnoli, M., Liguori, M., Simoni, S. and Guidi, S. (1998) Tomato as host plant: effect on some biological parameters of *Phytoseiulus persimilis* Athias-Henriot and *Tetranychus urticae* Koch strains. *Redia* 81, 93-99.

Castaldi, R. (1999) Possibility of biological control on aubergine. *Informatore Agrario* 55(46), 77-79.

Chant, D.A. (1985) Systematics and taxonomy. In: Helle, W. and Sabelis, M.W. (eds) *Spider Mites: Their Biology, Natural Enemies and Control*. Volume 1B. Elsevier, Amsterdam, pp. 17-29.

Chant, D.A. (1992) Trends in the discovery of new species and adult setal patterns in the family Phytoseiidae (Acari: Gamasina). *International Journal of Acarology* 18, 323-362.

Chant, D.A. and McMurtry, J.A. (1994) A review of the subfamilies Phytoseiinae and Typhlodrominae (Acari: Phytoseiidae). *International Journal of Acarology* 20, 233-310.

Cheng, Z., Horn, D.J., Lindquist, R.K. and Taylor, R.A.J. (1996) Fuzzy analysis for a greenhouse spider mite management system. *Ecological Modelling* 90, 111-121.

Chermiti, B. (1992) Biological control II: biological control of an artificial infestation of *Tetranychus urticae* Koch. using *Phytoseiulus persimilis* Athias-Henriot in a protected aubergine crop. *Mededelingen van de Faculteit Landbouwwetenschappen, Rijksuniversiteit Gent* 57, 959-964.

Cloutier, C., Arodokoun, D., Johnson, S.G. and Gelinas, L. (1995) Thermal dependence of *Amblyseius cucumeris* (Acarina: Phytoseiidae) and *Orius insidiosus* (Heteroptera: Anthocoridae) in greenhouses. In: Parker, B.L., Skinner, M. and Lewis, T. (eds) *Thrips Biology and Management: Proceedings of the 1993 International Conference on Thysanoptera*. Plenum Publishing Co. Ltd., London, pp. 231-235.

Collyer, E. (1982) The Phytoseidae of New Zealand (Acarina). 1. The genera *Typhlodromus* and *Amblyseius* - keys and new species. *New Zealand Journal of Zoology* 9, 185-206.

Courcy Williams, M. de (1993) *Amblyseius cucumeris* (Acari: Phytoseiidae) in the control of western flower thrips (*Frankliniella occidentalis*) on cyclamen. *Bulletin OILB/SROP* 16, 48-52.

Croft, B.A., Monetti, L.N. and Pratt, P.D. (1998) Comparative life histories and predation types - are *Neoseiulus Californicus* and *N. fallacis* (Acari, Phytoseiidae) similar type II selective predators of spider mites. *Environmental Entomology* 27(3), 531-538.

Dissevelt, M., Altena, K. and Ravensberg, W.J. (1995) Comparison of different Orius species for control of *Frankliniella occidentalis* in glasshouse vegetable crops in The Netherlands. *Mededelingen Faculteit Landbouwkundige en Toegepaste Biologische Wetenschappen, Universiteit Gent* 60(3a), 839-845.

Dong, H.F., Guo, Y.J. and Niu, L.P. (1986) Biological control of the two spotted spider mites with *Phytoseiulus persimilis* on four ornamental plants in greenhouses. *Chinese Journal of Biological Control* 2, 59-62.

Drukker, B., Janssen, A., Ravensberg, W. and Sabelis, M.W. (1997) Improved control capacity of the mite predator *Phytoseiulus persimilis* (Acari: Phytoseiidae) on tomato. *Experimental & Applied Acarology* 21, 507-518.

Ehara, S. and Amano, H. (1993) *Neoseiulus californicus* (McGregor). In: Ehara, S. (ed.) *Plant Mites of Japan in Colors*. Zenkoku Noson Kyoiku Kyokai, Tokyo, pp. 10-11.

El-Laithy, A.Y. (1996) Integrated control of two spotted spider mite, *Tetranychus urticae* on cucumber grown under plastic house conditions in Egypt. *Entomophaga* 41, 485-491.

El-Laithy, A.Y.M. and Elsawi, S.A. (1998) Biology and life table parameters of the predatory mite *Neoseiulus californicus* fed on different diet. *Zeitschrift fur Pflanzenkrankheiten und Pflanzenschutz* 105(5), 532-537.

Field, R.P. and Hoy, M.A. (1986) Evaluation of genetically improved strains of *Metaseiulus occidentalis* (Nesbitt) (Acarina: Phytoseiidae) for integrated control of spider mites on roses in greenhouses. *Hilgardia* 54(2), 1-32.

Fouly, A.H. and El-Laithy, A.Y.M. (1992) Immature stages and life history of the predatory mite species *Amblyseius barkeri* (Hughes,1948) (Acarina, Gamasida, Phytoseiidae). *Deutsche Entomologische Zeitschrift* 39, 427-435.

Friese, D.D. and Gilstrap, F.E. (1985) Prey requirements and developmental times of three phytoseiid. species predaceous on spider mites. *Southwestern Entomologist* 10, 83-88.

Gillespie, D.R. (1989) Biological control of thrips (Thysanoptera: Thripidae) on greenhouse cucumber by *Amblyseius cucumeris*. *Entomophaga* 34, 185-192.

Gillespie, D.R. and Quiring, D.J.M. (1994) Reproduction and longevity of the predatory mite, *Phytoseiulus persimilis* (Acari: Phytoseiidae) and its prey, *Tetranychus urticae* (Acari: Tetranychidae) on different host plants. *Journal of the Entomological Society of British Columbia* 91, 3-8.

Gillespie, D.R. and Ramey, C.A. (1988) Life history and cold storage of *Amblyseius cucumeris* (Acarina: Phytoseiidae). *Journal of Entomological Society of British Columbia* 85, 71-76.

Golovkina, L.S. and Zvereva, Yu.F. (1991) Resistant forms of *Phytoseiulus* as a component in the system of cucumber protection under greenhouse conditions. In: *Ekologicheskie Osnovy Primeneniya Insektoakaritsidov*. VIZR, St. Petersburg, Russia, pp. 49-55.

Gould, H.J., Hussey, N.W. and Parr, N.J. (1969) Large scale control of *Tetranychus urticae* Koch on cucumbers by the predator *Phytoseiulus persimilis* A.H. In: Evans, G.O. (ed.) *Proceedings of the 2nd International Congress of Acarology*, Sutton Bonington, England, 19-25 July, 1967. Akademiai Kiado, Budapest, pp. 383-388.

Grafton-Cardwel, E.E., Ouyang, Y. and Striggow, R.A. (1999) Predacious mites for control of citrus thrips, *Scirtothrips citri* (Thysanoptera: Thripidae) in nursery citrus. *Biological Control* 14, 29-36.

Grasselly, D., Millot, P. and Alauzet, C. (1995) Biological control of the thrips *Frankliniella occidentalis* on cucumber. *Infos Paris* 110, 34-37.

Guo, Y.J. and Dong, H.F. (1987) Effect of fluctuating temperature and relative humidity on the development and survival of *Phytoseiulus persimilis*. *Chinese Journal of Biological Control* 3, 19-22.

Hamlen, A.P. and Poole, R.T. (1980) Effects of a predacious mite on spider mite populations of *Diffenbachia* under greenhouse and interior environments. *HortScience* 15, 611-612.

Hariyappa, A.S. and Kulkarni, K.A. (1988) Biology and feeding efficiency of the predatory mite, *Amblyseius longispinosus* (Evans) and chilli mite, *Polyphagotarsonemus latus* (Banks). *Journal of Biological Control* 2, 131-132.

Helle, W. and Sabelis, M.W. (eds) (1985) *Spider Mites: Their Biology, Natural Enemies and Control*. Volume 1B. Elsevier, Amsterdam. 458 pp.

Hessein, N.A. and Parrella, M.P. (1990) Predatory mites help control thrips on floriculture crops. *California Agriculture* 44 (6), 19-21.

Higgins, C.J. (1992) Western flower thrips (Thysanoptera: Thripidae) in greenhouses: population dynamics, distribution on plants, and associations with predators. *Journal of Economic Entomology* 85, 1891-1903.

Ho, C.C., Lo, K.C. and Chen, W.H. (1995) Comparative biology, reproductive compatibility, and geographical distribution of *Amblyseius longispinosus* and *A. womersleyi* (Acari: Phytoseiidae). *Environmental Entomology* 24, 601-607.

Houten, Y.M. van (1996) Biological control of western flower thrips on cucumber using the predatory mites *Amblyseius cucumeris* and *A. limonicus*. *Bulletin OILB-SROP* 19(1), 59-62.

Houten, Y.M. van and Stratum, P. van (1993) Biological control of western flower thrips in greenhouse sweet peppers using nondiapausing predatory mites. *Proceedings of the Section Experimental and Applied Entomology of The Netherlands Entomological Society* 4, 229-234.

Houten, Y.M. van, Stratum, P. van, Bruin, J. and Veerman, A. (1995) Selection for non-diapause in *Amblyseius cucumeris* and *Amblyseius barkeri* and exploration of the effectiveness of selected strains for thrips control. *Entomologia Experimentalis et Applicata* 77, 289-295.

Hughes, A.M. (1976) *Mites of Stored Food and Houses. Technical Bulletin of the Ministry of Agriculture, Fisheries and Food, No. 9.* HMSO, London. 400 pp.

Hussey, N.W. and Huffaker, L.B. (1976) Spider mites. In: Delucchi, V.L. (ed.) *Studies in Biological Control*. Cambridge University Press, Cambridge, pp. 179-228.

Hussey, N.W. and Scope, N.E.A. (1985) *Biological Pest Control. The Glasshouse Experience*. Blandford Press, Dorset, 240 pp.

Jacobson, R.J. (1995) Integrated pest management in cucumbers - prevention of establishment of *Frankliniella occidentalis* (Pergande). *Mededelingen Faculteit Landbouwkundige en Toegepaste Biologische Wetenschappen, Universiteit Gent* 60(3a), 857-863.

Jarosik, V. and Pliva, J. (1995) Assessment of *Amblyseius barkeri* (Acarina: Phytoseiidae) as a control agent for thrips on greenhouse cucumbers. *Acta Societatis Zoologicae Bohemicae* 59(3-4), 177-186.

Kazak, C., Colkesen, T., Karut, K. and Sekeroglu, E. (1997) Biological control of *Tetranychus cinnabarinus* by *Phytoseiulus persimilis* on greenhouse cultivars. *Bulletin OILB-SROP* 20(4), 215-220.

Kazak, C., Karut, K., Yildiz, S. and Sekeroglu, E. (1999) Biological control of *Thrips tabaci* Lindeman (Thysanoptera: Thripidae) by *Amblyseius cucumeris* (Oudemans) (Acarina: Phytoseiidae) on greenhouse cucumber in Adana, Turkey. *Acta Horticulturae* 492, 169-174.

Klashorst, G. van de, Readshaw, J.L., Sabelis, M.W. and Lingeman, R. (1992) A demonstration of asynchronous local cycles in an acarine predator-prey system. *Experimental & Applied Acarology* 14, 185-199.

Kropezynska, D. and Tomczyk, A. (1996) Development of *Tetranychus urticae* Koch and *Tetranychus cinnabarinus* Boisd, populations on sweet pepper and *Phytoseiulus persimilis* (A.-H) effectiveness in their control. *Bulletin OILB SROP*

19(1), 71-74.

Kurogi, S., Nakamura, M. and Kawasaki, Y. (1997) Studies on integrated control of major insect pests of sweet pepper in a greenhouse [in Japan]. 3. Control of *Thrips palmi* with 2 species of predators, *Orius sauteri* and *Amblyseius cucumeris*. *Proceedings of the Association for Plant Protection of Kyushu* 43, 106-109.

Kwon, G.M., Lee, Y.I., Lee, S.W. and Choi, K.H. (1998) Development and prey consumption of phytoseiid mites, *Amblyseius womersleyi, A. fallacis,* and *Typhlodromus occidentalis* under controlled environments. *Korean Journal of Applied Entomology* 37, 53-58.

Lindquist, R.K., Casey, M.L., Bauerle, W.L. and Short, T.L. (1987) Effects of an overhead misting system on thrips populations and spider mite-predator interactions on greenhouse cucumber. *Bulletin SROP* 10, 97-100.

Loginova, E., Atanassov, N. and Georgiev, G. (1987) Biological control of pests and diseases in glasshouses in Bulgaria - today and in the future. *Bulletin SROP* 10, 101-105.

Markkula, M., Tiittanen, K. and Hokkanen, H.M.T. (1987) Failures in biological control of spider mites - due to predatory mites or their users? *Bulletin SROP* 10(2), 108-110

McClanahan, R.J. (1968) Influence of temperature of the reproductive potential of two mite predators of the twospotted spider mite. *Canadian Entomologist* 100, 549-556.

McMurtry, J.A. and Croft, B.A. (1997) Life-styles of phytoseiid mites and their roles in biological control. *Annual Review of Entomology* 42, 291-321.

Momen, F.M. (1995) Feeding, development and reproduction of *Amblyseius barkeri* (Acarina: Phytoseiidae) on various kinds of food substances. *Acarologia* 36, 101-105.

Momen, F.M. (1996) Effect of prey density on reproduction, prey consumption and sex-ratio of *Amblyseius barkeri* (Acari: Phytoseiidae). *Acarologia* 37, 1-6.

Moraes, G.J. de, McMurtry, J.A. and Denmark, H.A. (1986) *A Catalog of the Mite Family Phytoseiidae. References to Taxonomy, Synonymy, Distribution and Habitats.* EMBRAPA, Brazil, 353 pp.

Morewood, W.D. and Gilkeson, L.A. (1991) Diapause induction in the thrips predator *Amblyseius cucumeris* (Acarina: Phytoseiidae) under greenhouse conditions. *Entomophaga* 36, 253-263.

Nakao, H., Saito, Y. and Mori, H. (1987) Biological control of the two-spotted spider mite by the resistant strain of *Phytoseiulus persimilis*. I. Estimation of effectiveness of the predator by experiments and simulation under chemical spraying conditions (Acari, Phytoseiidae; Tetranychidae). *Japanese Journal of Applied Entomology and Zoology* 31, 359-368.

Nawar, M.S. and El-Sherif, A.A. (1993) *Neoseiulus cucumeris* (Oudemans), a predator of whitefly *Bemisia tabaci* (Gennadius). *Bulletin of the Entomological Society of Egypt* 71, 9-17.

Nihoul, P. (1993a) Asynchronous populations of *Phytoseiulus persimilis* Athias-Henriot and effective control of *Tetranychus urticae* Koch on tomatoes under glass. *Journal of Horticultural Science* 68, 581-588.

Nihoul, P. (1993b) Controlling glasshouse climate influences the interaction between tomato glandular trichome, spider mite and predatory mite. *Crop Protection* 12, 443-447.

Nihoul, P. (1994) Phenology of glandular trichomes related to entrapment of *Phytoseiulus persimilis* A.-H. in the glasshouse tomato. *Journal of Horticultural Science* 69, 783-789.

Nihoul, P. and Hance, T. (1993) Use of a damage index to evaluate the biological control of the two-spotted spider mite *Tetranychus urticae* Koch (Acari; Tetranychidae) on tomato crops. *Journal of Horticultural Science* 68, 575-580.

Nomikou, M., Janssen, A., Schraag, R. and Sabelis, M.W. (2001) Phytoseiid predators as potential biological control agents for *Bemisia tabaci*. *Experimental & Applied Acarology* 25, 271-291.

Oncuer, C., Yoldas, Z., Madanlar, N. and Gul, A. (1994) Biological control of insect pests on greenhouse vegetables in the Aegean Region. In: *Turkiye III. Biyolojik Mucadele Kongresi Bildirileeri, 25-28 Ocak 1994, Ege Universitesi Ziraat Fakultesi, Bitki Koruma Bolumu, Izmir.* Ege Universitesi Basimevi, Bornova, Turkey, pp. 395-407.

Palevsky, E., Reuveny, H., Okonis, O. and Gerson, U. (1999) Comparative behavioural studies of larval and adult stages of the phytoseiids (Acari: Mesostigmata) *Typhlodromus athiasae* and *Neoseiulus californicus*. *Experimental & Applied Acarology* 23(6), 467-485.

Parr, W.J. (1969) Red spider mite *Tetranychus urticae*. *Report of the Glasshouse Crops Research Institute* 1969-1970, 108-109.

Parr, W.J. and Hussey, N.W. (1966) Diapause in the glasshouse red spider mite *Tetranychus urticae* Koch: a synthesis of present knowledge. *Horticultural Research* 6, 1-21.

Pruszynski, S., Siwek, L., Aumiller, P. and Konopinska, M. (1985) Results of biological control of spider mites (Tetranychidae) using the predacious mite *Phytoseiulus persimilis* in production glasshouses of the Naramowice state horticultural farm. *Prace Naukowe Instytutu Ochrony Roslin* 24, 161-172.

Ramakers, P.M.J. (1988) Population dynamics of the thrips predators *Amblyseius mckenziei* and *Amblyseius cucumeris* (Acarina: Phytoseiidae) on sweet pepper. *Netherlands Journal of Agricultural Science* 36, 247-252.

Ramakers, P.M.J., Dissevelt, M. and Peeters, K. (1989) Large scale introductions of phytoseiid predators to control thrips on cucumber. *Mededelingen van de Faculteit Landbouwwetenschappen, Rijksuniversiteit Gent* 54 (3a), 923-929.

Ramakers, P.M.J. and Voet, S.J.P. (1996) Use of castor bean, *Ricinus communis*, for the introduction of the thrips predator *Amblyseius degenerans* on glasshouse-grown sweet peppers. *Mededelingen van de Faculteit Landbouwwetenschappen, Rijksuniversiteit Gent* 60: 3a, 885-891.

Rijn, P.C.J. van and Sabelis, M.W. (1990) Pollen availability and its effect on the maintenance of populations of *Amblyseius cucumeris*, a predator of thrips. *Mededelingen van de Faculteit Landbouwwetenschappen, Rijksuniversiteit Gent* 55, 2a, 335-341

Rijn, P.C.J. van and Sabelis, M.W. (1993) Does alternative food always enhance biological control? The effect of pollen on the interaction between western flower thrips and its predators. *Bulletin OILB/SROP* 16 (8), 123-125.

Rodriguez-Reina, J.M., Ferragut, F., Carnero, A. and Pena, M.A. (1994) Diapause in the predacious mites *Amblyseius cucumeris* (Oud.) and *Amblyseius barkeri* (Hug.): consequences of use in integrated control programmes. *Journal of Applied Entomology* 118, 44-50.

Saito, Y., Urano, S., Nakao, H., Amimoto, K. and Mori, H. (1996) A simulation model for predicting the efficiency of biological control of spider mites by phytoseiid predators. 2. Validity tests and data necessary for practical usage. *Japanese Journal of Applied Entomology and Zoology* 40, 113-120.

Sanchez, J.A., Garcia, F., Lacasa, A., Gutierrez, L., Oncina, M., Contreras, J. and Gomez, Y.J. (1997) Response of the anthocorids *Orius laevigatus* and *Orius albidipennis* and the phytoseiid *Amblyseius cucumeris* for the control of *Frankliniella occidentalis* in commercial crops of sweet peppers in plastic houses in Murcia, Spain. *Bulletin OILB-SROP* 20(4), 177-185.

Schausberger, P. and Walzer, A. (2001) Combined versus single species release of predaceous mites: predator-predator interactions and pest suppression. *Biological Control* 20(3), 269-278.

Schelt, J. van and Altena, K. (1997) Growing tomatoes without insecticides. *Proceedings of the Section Experimental and Applied Entomology of The Netherlands Entomological Society* 8, 151-157.

Simmonds, S.P. (1972) Observations on the control of *Tetranychus urticae* on roses by *Phytoseiulus persimilis*. *Plant Pathology* 21, 163-165.

Smith, V.A., Orr, D.B. and Hart, E.R. (1993) Economic analysis of two-spotted spider mite management on greenhouse-grown poplars. *Tree Planters' Notes* 44(4), 154-156.

Stanton, G. (1994) Thrips management and biological control. *GrowerTalks* 58, 36-40.

Takafugi, A. and Chant, D.A. (1976) Comparative studies of two species of predacious phytoseiid mites (Acarina: Phytoseiidae), with special reference to their responses to the density of their prey. *Researches on Population Ecology* 17, 255-310.

Takahashi, F. and Chant, D.A. (1992) Adaptive strategies in the genus *Phytoseiulus* Evans (Acari: Phytoseiidae). I. Development times. *International Journal of Acarology* 18, 171-176.

Takahashi, F. and Chant, D.A. (1993a) Phylogenetic relationships in the genus *Phytoseiulus* Evans (Acari: Phytoseiidae). I. Geographical distribution. *International Journal of Acarology* 19, 15-22.

Takahashi, F. and Chant, D.A. (1993b) Phylogenetic relationships in the genus *Phytoseiulus* Evans (Acari: Phytoseiidae). II. Taxonomic review. *International Journal of Acarology* 19, 23-37.

Tomczyk, A., Pilko, A. and Lenteren, J.C. van (1996) Prospects for integration of the defence abilities of the host plant and Phytoseiulus persimilis activity in spider mite control on cucumber. *Bulletin OILB-SROP* 19 (1), 175-178.

Toyashima, S. and Amano, H. (1998) Effect of prey density on sex ratio of two predacious mites, *Phytoseiulus persimilis* and *Amblyseius womersleyi* (Acari: Phytoseiidae). *Experimental and Applied Acarology* 22, 709-723.

Vacante, V. and Firullo, V. (1983) Observations on the population dynamics of *Phystoseiulus persimilis* A. H. (Acarina: Phytoseiidae) on roses in cold greenhouses in the Ragusa Province in Sicily. *Mededelingen van de Faculteit Landbouwwetenschappen, Rijksuniversiteit Gent* 48, 263-272.

Vacante, V. and Garzia, G.T. (1987) Present possibilities of control against *Tetranychus urticae* Koch by means of *Phytoseiulus persimilis* Athias-Henriot on rose and gerbera in a protected environment. *Difesa delle Piante* 10, 139-147.

Vacante, V. and Garzia, G.T. (1993) Programmed releases of *Amblyseius cucumeris* (Oudemans) against *Frankliniella occidentalis* (Pergande) on sweet pepper in unheated greenhouses. *Colture Protette* 22(1), 23-32.

Wardlow, L.R., Tobin, A.S. and Richards, S.G. (1991) Integrated pest management in year-round chrysanthemums during the winter months. *Mededelingen van de Faculteit Landbouwwetenschappen, Rijksuniversiteit Gent* 56 (2a), 259-264.

Workman, P.J., Dymock, J.J., Martin, N.A. and Ennis, I.L. (1994) The potential for biological control of thrips in greenhouse capsicums using the predatory mite, *Amblyseius cucumeris*. In: Popay, A.J. (ed.) *Proceedings of the Forty Seventh New Zealand Plant Protection Conference, Waitangi Hotel, New Zealand, 9-11 August, 1994*. New Zealand Plant Protection Society, Rotorua, New Zealand, pp. 139-143.

Workman, P.J. and Martin, N.A. (2000) Movement of *Phytoseiulus persimilis* (Acari: Phytoseiidae) on the leaves of greenhouse carnations and other cut flowers. *New Zealand Journal of Crop and Horticultural Science* 28, 9-15.

Yoldas, Z., Madanlar, N., Gul, A. and Onogur, E. (1999) Investigations on integrated control practices in vegetable glasshouses in Izmir. *Acta Horticulturae* 491, 453-460.

Zacharda, M. and Hluchy, M. (1996) Biological control of two-spotted spider mite *Tetranychus urticae* on strawberries by the predatory phytoseiid mite *Typhlodromus pyri*. *Bulletin OILB-SROP* 19(4), 226-230.

Zhang, Z.-Q. (1995) Variance and covariance of ovipositional rates and developmental rates in the Phytoseiidae (Acari: Mesostigmata): a phylogenetic consideration. *Experimental & Applied Acarology* 19, 139-146.

Zhang, Z.-Q. and Croft, B.A. (1994) A comparative life history study of immature *Amblyseius fallacis*, *Amblyseius andersoni*, *Typhlodromus occidentalis* and *Typhlodromus pyri* (Acari: Phytoseiidae) with a review of larval feeding patterns in the family. *Experimental & Applied Acarology* 18, 635- 657.

Zhang, Z.-Q. and Sanderson, J.P. (1995) Twospotted spider mites (Acari: Tetranychidae) and *Phytoseiulus persimilis* (Acari: Phytoseiidae) on greenhouse roses: spatial distribution and predator efficacy. *Journal of Economic Entomology* 88, 352-357.

Laelapid Mites 11

11.1 Introduction

The family Laelapidae is a member of the superfamily Dermanyssoidea in the order Mesostigmata.

Mites of this family include many vertebrate parasites, some of which attack domestic animals and are of veterinary importance. Species of *Hypoaspis* are free-living predators common in soil and have been employed in the biological control of soil-inhabiting mites and insects in greenhouses.

The Laelapidae is a large family of worldwide distribution. The taxonomy of this group is in need of serious revision. Members of the subfamily Hypoaspinae are commonly found in soil and litter, stored products, in the nests of vertebrates and arthropods, and phoretic on insects. Many species of the genus *Hypoaspis* are generalist predators.

References & suggested further reading. Hughes (1976).

11.2 Morphological characters

Laelapids are well-sclerotized mites of medium to large size. They are often brownish in colour when alive.

The chelicerae have prominent dentate digits, with a pilus dentilis on the fixed digit. The deutosternum has five to seven transverse rows of denticles. The dorsal shield is entire. The presternal area is often reticulate. The sternal shield of the female has three (rarely four) pairs of setae and two pairs of pores. The genital shield is rounded posteriorly. The male has a holoventral shield, rarely with separate sternal, genital and anal shields. Genu I and tibia I normally have three ventral setae.

References & suggested further reading. Hughes (1976); Deng *et al.* (1993).

11.3 Life history and biology

The life cycle consists of egg, larva, protonymph, deutonymph and adult stages. Many species are parthenogenetic. Reproductive rates of the laelapids are generally lower than that of Phytoseidae.

Most of the biological data of this family concern parasitic species and are not relevant to this book. Members of *Hypoaspis* feed on a variety of small arthropods and nematodes. They can complete egg to adult development in one to two weeks depending on temperature. Adults are cannibalistic, mostly attacking eggs.

11.4 Species important in greenhouses

Only two species, *Hypoaspis aculeifer* (Canestrini) and *Hypoaspis miles* (Berlese) have been used in greenhouses to control a number of soil inhabiting pest insects and mites. These two species can be easily separated: the dorsal shield of *H. aculeifer* has 39 setiform setae (Fig. 11.1A), whereas that of *H. miles* has 37 narrowly leaf-shaped setae (Fig. 11.2A and B).

11.4.1 *Hypoaspis aculeifer* (Canestrini)

11.4.1.1 Common Name

No common names have been used for this species, which is also known as *Gaeolaelaps aculeifer.*

11.4.1.2 Distribution And Prey

This mite is a polyphagous predator, feeding on other mites, small insects and nematodes. It is widely distributed in Europe, Asia and North America. It is a common member of the soil fauna and is rarely found on aerial parts of the plants.

References & suggested further reading. Hughes (1976); Deng *et al.* (1993).

11.4.1.3 Appearance And Diagnostic Characters

This is medium-sized laelapid species, with adult females measuring 600-940 μm. It is brown in colour when alive. The dorsal shield is well sclerotized, bearing 39 pairs of setae (Fig. 11.1A) and covered with faint reticulation. All dorsal setae are setiform (cf: narrow leaf-shaped in *H. miles*). The genital shield is flask-shaped. The metasternal seta and associated pores are on interscutal membrane.

References & suggested further reading. Hughes (1976); Deng *et al.* (1993).

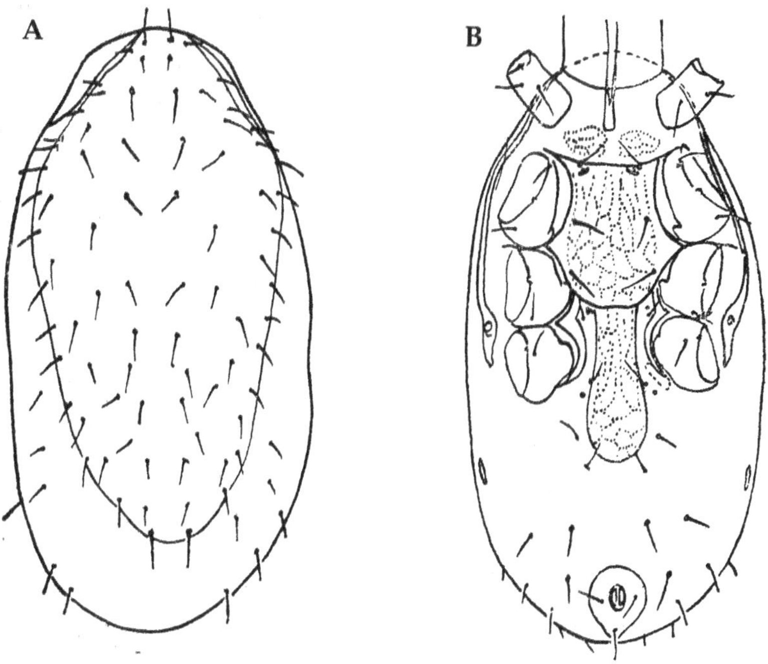

Fig. 11.1 ***Hypoaspis aculeifer*** (Canestrini), adult females. A, dorsal view; B, ventral view (after Deng *et al.*, 1993).

11.4.1.4 LIFE HISTORY AND BIOLOGY

The developmental time varies with types of food and temperatures. The egg to adult development takes ten to 13 days at 26°C with *Tyrophagus putrescentiae* as food. With larvae of *R. robini* as prey, this species reaches adulthood in 11-12 days and females lay an average of 114 eggs during an oviposition period of 41 days.

Reproduction is arrhenotokous. Sex ratio is slightly female-biased (1.1 female vs 1.0 male). The preoviposition period is much shorter and the oviposition period much longer in female-only populations than in bisexual populations, and fecundity in female-only populations is just over half of those in sexual populations.

Larvae of this species do not feed. Starved adults of *H. aculeifer* may feed on their eggs, and females on males.

References & suggested further reading. Kevan and Sharma (1964); Barker (1968); Ignatowicz (1974); Usher and Davis (1983); Ragusa *et al.* (1986); Sardar and Murphy (1987); Ragusa and Zedan (1988); Zedan (1988); Murphy and Sardar (1991); Ruf (1991).

11.4.1.5 Use In Biocontrol

This species is an effective biological control agent against *Rhizoglyphus* bulb mites (see 8.4.1.6 and 8.4.2.6). It can also be released to control sciarid flies (*Bradysia* spp.) in growing media of crops such as poinsettia in greenhouses. It can also be released to reduce densities of thrips pupae and other acarid mites (*Tyrophaus*) in the soil or growing media.

References & suggested further reading. Conijn *et al.* (1997); Piatkowski (1997); Zedan (1988); Lesna *et al.* (1995, 1996, 2000).

11.4.2 *Hypoaspis miles* (Berlese)

11.4.2.1 Common Name

No common names, but also known as *Stratiolaelaps miles*.

11.4.2.2 Distribution And Prey

This mite is a polyphagous species and is widely distributed in Europe, Asia and North America. It is a common member of the soil fauna and feeds on small insects, mites and nematodes.

References & suggested further reading. Hughes (1976); Deng *et al.* (1993); Enkegaard *et al.* (1995).

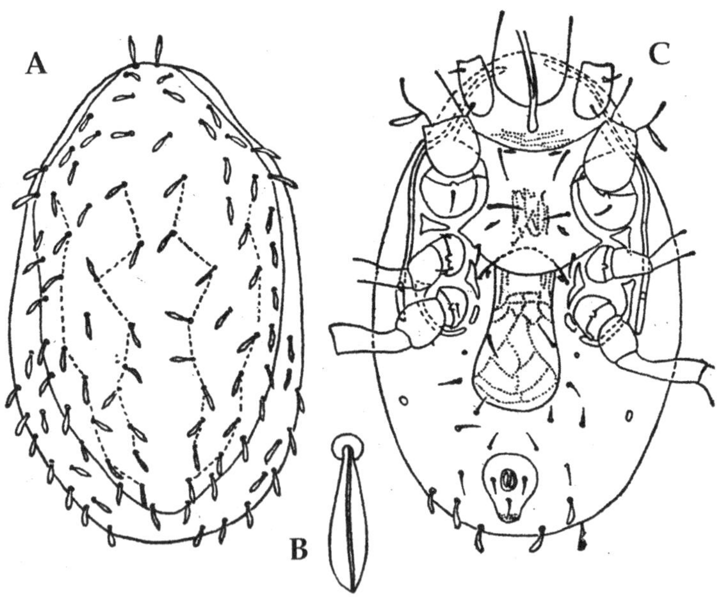

Fig. 11.2 *Hypoaspis miles* Berlese, adult females. A, dorsal view; B, dorsal seta, enlarged view; C, ventral view (after Deng *et al.*, 1993).

11.4.2.3 Appearance And Diagnostic Characters

This is medium-sized laelapid species, with adult females measuring about 650 μm. It is brown in colour when alive. The dorsal shield is well sclerotized, bearing 37 pairs of setae (Fig. 11.2A) and reticulations. All dorsal setae are narrow leaf-shaped (cf: setiform in *H. aculeifer*). The sternal shield extends to mid-coxa III and has three pairs of setae and two pairs of pores.

References & suggested further reading. Hughes (1976); Deng *et al.* (1993).

11.4.2.4 Life History And Biology

Developmental time from the egg to adult varies with temperature and types of food and takes 34 days at 15°C but nine days at 28°C when feeding on *Bradysia* larvae. The developmental threshold temperature is between 10 and 12°C. Females lay an average of 44 eggs in 53 days when feeding on larvae of *Lycoriella* and 22 eggs in 69 days when feeding by *Tyrophagus*. Males live twice as long as females.

This is also an arrhenotokous species and only unfertilized eggs develop into males. The sex ratio is slightly female-biased, being 66% females when feeding on *Lycoriella* larvae and 54% females when feeding on *Tyrophagus* mites. When fed *Acarus siro*, females lay two to three eggs/day and with food, 60% of males and females can survive for 142 days. When fed immature stages of *Rhizoglyphus robini* at 28±1°C, the duration of development of the larva, protonymph and deutonymph is 1.2, 2.1 and 2.4 days, respectively. The larvae do not feed.

References & suggested further reading. Hoda *et al.* (1987); Rasmy *et al* (1987); Matteoni *et al.* (1993); Wright and Chambers (1994); Brodsgaard *et al.* (1996); Enkegaard *et al.* (1997); Ydergaard *et al.* (1997).

11.4.2.5 Use In Biocontrol

Hypoaspis miles is an effective predator of sciarid larvae. When released at 55 mites per pot, it provides satisfactory control of *Bradysia* in pot-grown *Cyclamen* and poinsettias in greenhouses. *H. miles* are found mostly in the top 1 cm of compost and can persist in the compost for up to seven weeks in the absence of *Bradysia*.

This species also feeds on thrips pupae, shore fly larvae (*Scatella*), and acarid mites (*Rhizoglyphus* and *Tyrophagus*) and can be released to reduce densities of these pests in the soil or growing media.

References & suggested further reading. Chambers *et al.* (1993); Lindquist *et al.* (1994); Enkegaard *et al.* (1995).

11.4.3 Other laelapid species

A species of *Hypoaspis* near *H. aculeifer* can significantly reduce numbers of larvae and adults of *Bradysia* spp. over a ten-week period when released inundatively at a rate of 6,000 mites/plant to the sawdust substrate of hydroponically grown greenhouse cucumbers. An inoculative introduction of 125 mites/plant to cucumber plants reduces peak numbers of *Bradysia* spp. by about 20%. This species also feeds on thrips pupae and can reduce emergence of adult *Frankliniella occidentalis* by 30% when released at 1,600 mites/plant.

Another laelapid mite, *Hypoaspis vacua* (Michael), can develop successfully when feeding on nymphs of *R. echinopus*. Immature stages of *H. vacua* consume a total of 33 nymphs of *R. echinopus*, whereas adult females consume 13 nymphs daily.

References & suggested further reading. Abou-Awad *et al.* (1989); Gillespie and Quiring (1990).

11. 5 References cited

Abou-Awad, B.A., Nasr, A.K., Gomaa, E.A. and Abou-Elela, M.M. (1989) Feeding, development and reproduction of the predatory mite, *Hypoaspis vacua* on various kinds of food substances (Acari: Laelapidae). *Insect Science and its Application* 10, 503-506.

Barker, P.S. (1968) The response of a predator, *Hypoaspis aculeifer* (Canestrini) (Acarina: Laelapidae), to two species of prey. *Canadian Journal of Zoology* 47, 343-345.

Brodsgaard, H.F., Sardar, M.A., Enkegaard, A. and Lenteren, J.C. van (1996) Prey preference of *Hypoaspis miles* (Berlese) (Acarina: Hypoaspididae): non-interference with other beneficials in glasshouse crops. *Bulletin OILB-SROP* 19, 23-26.

Chambers, R.J., Wright, E.M. and Lind, R.J. (1993) Biological control of glasshouse sciarid flies (*Bradysia* spp.) with the predatory mite, *Hypoaspis miles*, on cyclamen and poinsettia. *Biocontrol Science and Technology* 3, 285-293

Conijn, C.G.M., Lesna, I. and Altena, K. (1997) Biological control of the bulb mite *Rhizoglyphus robini* by the predatory mite *Hypoaspis aculeifer* on lilies: implementation in practice. *Acta Horticulturae* 430, 619-624.

Deng, G.-F., Wang, D.-Q., Gu, Y.-M. and Meng, Y.-C. (1993) *Economic Insect Fauna of China Facs. 40 Acari: Dermanyssoidea*. Science Press, Beijing. 391 pp.

Enkegaard, A., Brodsgaard, H.F. and Sardar, M.A. (1995) *Hypoaspis miles* — a polyphagous predatory mite for control of soil-dwelling pests: biology and food preference. *SP Rapport Statens Planteavlsforsog* 4, 247-255.

Enkegaard, A., Sardar, M.A. and Brodsgaard, H.F. (1997) The predatory mite

Hypoaspis miles: biological and demographic characteristics on two prey species, the mushroom sciarid fly, *Lycoriella solani*, and the mould mite, *Tyrophagus putrescentiae*. *Entomologia Experimentalis et Applicata* 82, 135-146.

Gillespie, D.R and Quiring, D.M.J. (1990) Biological control of fungus gnats, *Bradysia* spp. (Diptera: Sciaridae), and Western flower thrips, *Frankliniella occidentalis* (Pergrande) (Thysanoptera: Thripidae), in greenhouses using a soil-dwelling predatory mite, *Geolaelaps* sp. nr. *aculeifer* (Canestrini) (Acari: Laelapidae). *Canadian Entomologist* 122, 975-983.

Hoda, F.M., Taha, H.A., Ibrahim, G.A. and El-Beheri, M.M. (1987) Biological observations on the predator mite, *Hypoaspis miles* Berlese (Acarina: Laelapidae). *Bulletin de la Societe Entomologique d'Egypte* 66, 103-106.

Hughes, A.M. (1976) *Mites of Stored Food and Houses. Technical Bulletin of the Ministry of Agriculture, Fisheries and Food, No. 9*. HMSO, London. 400 pp.

Kevan, D.E.Mc.E. and Sharma, G.D. (1964) Observations on the biology of *Hypoaspis aculeifer* (Canestrini 1884), apprently new to North America (Acarina: Mesostigmata: Laelapidae). *Acarologia* 6, 647-658.

Ignatowicz, S. (1974) Observations on the biology and development of *Hypoaspis aculeifer* (Canestrini, 1884) (Acarina: Gamasides). *Zoologica Poloniae* 24, 41-59

Lesna, I., Sabelis, M.W., Bolland, H.R. and Conijn, C.G.M. (1995) Candidate natural enemies for control of *Rhizoglyphus robini* Claparede (Acari: Astigmata) in lily bulbs: exploration in the field and pre-selection in the laboratory. *Experimental & Applied Acarology* 19, 655-669.

Lesna, I., Sabelis, M.W. and Conijn, C. (1996) Biological control of the bulb mite, *Rhizoglyphus robini*, by the predatory mite, *Hypoaspis aculeifer*, on lilies: predator-prey interactions at various spatial scales. *Journal of Applied Ecology* 33, 369-376.

Lesna, I., Conijn, C.G.M., Sabelis, M.W. and Straalen, N.M. van (2000) Biological control of the bulb mite, *Rhizoglyphus robini*, by the predatory mite, *Hypoaspis aculeifer*, on lilies: predator-prey dynamics in the soil, under greenhouse and field conditions. *Biocontrol Science and Technology* 10, 179-193.

Lindquist, R., Buxton, J. and Piatkowski, J. (1994) Biological control of sciarid flies and shore flies in glasshouses. *Proceedings Brighton Crop Protection Conference, Pests and Diseases Volume 3*. British Crop Protection Council, Bracknell,UK, pp. 1067-1072.

Matteoni, J., Elliot, D., Cahn, D., Ravensberg, W.J. and Dissevelt, M. (1993) Chemical effects on greenhouse biological control agents. *GrowerTalks* 57, 81-85

Murphy, P.W. and Sardar, M.A. (1991) Resource allocation and utilization contrasts in *Hypoaspis aculeifer* (Can.) and *Alliphis halleri* (G. & R. Can.) (Mesostigmata) with emphasis on food source. In: Schuster, R. and Murphy, P.W. (eds) *The Acari. Reproduction, Development and Life-history Strategies*, Chapman & Hall, London, UK, pp. 301-311.

Piatkowski, J. (1997) Biological control of sciarid flies (*Bradysia* spp.) with predatory mites *Hypoaspis aculeifer* on poinsettia crops in greenhouses. *Bulletin OILB-SROP* 20(4), 221-224.

Ragusa, S. and Zedan, M.A. (1988) Biology and predation of *Hypoaspis aculeifer* (Canestrini) (Parasitiformes, Dermanyssidae) on *Rhizoglyphus echinopus* (Fum. & Rob.) (Acariformes, Acaridae). *Redia* 76, 213-126.

Ragusa, S., Zedan, M.A. and Sciacchitano, M.A. (1986) The effects of food from plant and animal sources on the development and egg production of the predaceous mite *Hypoaspis aculeifer* (Canestrini) (Parasitiformes, Dermanyssidae). *Redia* 69, 481-488.

Rasmy, A.H., Nasr, A.K. and Reda, A.S. (1987) Reproductive response and development of three soil predaceous mites utilizing the acarid mite *Tyrophagus casei* Oud. as an alternate diet. *Anzeiger fur Schadlingskunde, Pflanzenschutz, Umweltschutz* 60, 92-94.

Ruf, A. (1991) Do females eat males? Laboratory studies on the population development of *Hypoaspis aculeifer* (Acari: Parasitiformes). In: Dusbabek, F. and Bukva, V. (eds) *Modern Acarology. Volume II. Proceedings of the VIII International Congress of Acarology,* held in Ceske Budejovice, Czechoslovakia, 6-11 August 1990. SPB Academic Publishing bv., The Hague, The Netherlands, pp. 487-492.

Sardar, M.A. and Murphy, P.W. (1987) Feeding tests of grassland soil-inhabiting gamasine predators. *Acarologia* 28, 117-121.

Wright, E.M. and Chambers, R.J. (1994) The biology of the predatory mite *Hypoaspis miles* (Acari: Laelapidae), a potential biological control agent of *Bradysia paupera* (Dipt.: Sciaridae). *Entomophaga* 39, 225-235.

Usher, M.B. and Davis, P.R. (1983) The biology of *Hypoaspis aculiefer* (Canestrini) (Mesostigmata): is there a tendency towards social behaviour? *Acarologia* 24, 243-250.

Ydergaard, S., Enkegaard, A. and Brodsgaard, H.F. (1997) The predatory mite *Hypoaspis miles*: temperature dependent life table characteristics on a diet of sciarid larvae, *Bradysia paupera* and *B. tritici*. *Entomologia Experimentalis et Applicata* 85, 177-187.

Zedan, M.A.A. (1988) Studies on predator-prey interactions between *Hypoaspis aculeifer* Canestrini (Acarina: Laelapidae) and *Rhizoglyphus echinopus* (Fum. & Rob.) (Acarina: Acaridae) under laboratory conditions. *Revue de Zoologie Africaine* 102, 381-387.

Other Beneficial Mites 12

12.1 Introduction

Major groups of beneficial mites that are commonly used for biological control in greenhouses have been covered in the last two chapters. Several other families occasionally occur in greenhouses or have potential importance as biological agents against greenhouse pests. These families are introduced here in this chapter.

12.2 Mesostigmata

12.2.1 Parasitidae

The Parasitidae is the only family of the superfamily Parasitoidea. They are common predators of the soil fauna and are distributed worldwide.

The parasitids are medium to large predatory mites, often yellowish to dark brown in colour. The checlicerae are strong and dentate. The males have a spermatotreme on the movable digit. The hypostome bears four pairs of subcapitular setae and ten or more rows of denticles. The idiosoma is covered by a large dorsal shield or two shields, with more dorsal setae on the posterior half. The sternal shield in the female bears three pairs of setae. The fourth pair of sternal setae are on a pair of large metasternal shield, which flanks the anterolateral margins of a triangular genital shield (Fig. 2.13.1A). The male genital opening is presternal and the genital shield overlies the reduced bases of the tritosternum. The ventrianal shield is often fused with the podal, peritrematic and, more rarely, dorsal shield. The legs are long, terminating in a pair of claws, which may be absent in leg I.

The family consists of subfamilies Pergamasinae and Parasitinae. Members of the former are mainly found in the soil, whereas those of the latter are more common in temporary accumulations of organic matter and in nests of animals. Dispersal is via deutonymphs phoretic on insects.

Parasitus fimetorum (Berlese) is commonly found in association with bulb mites in soil and can suppress *Rhizoglyphus robini* on lily bulb propagation when peat is used as the growing medium.

Parasitus bituberosus Karg feeds on eggs and first and second instars of *Lycoriella* flies, Collembola and other insects and nematodes in mushroom compost. It has good potential in the biological control of sciarid flies and midges in compost and should be explored for control of sciarid flies in greenhouses.

Several species of *Pergamasus* feed on *Tyrophagus* in the soil. Their effectiveness is unknown and should be investigated in the future.

References & suggested further reading. Karg (1961); Al-Amidi and Downes (1990); Al-Amidi *et al.* (1991); Evans (1992); Lesna *et al.* (1995).

12.2.2 Ascidae

The Ascidae is family of the superfamily Ascoidea. They are predatory mites in soil, on plants or in association with other animals.

The ascids are small to medium in size (most species 300-500 μm), and are often pale, yellow to brown in colour. The chelicerae are dentate, serrate, or rarely with regressed fixed digit. The palps have six setae on the genu and a two-tined apotele on the tarsus. The idiosoma has one single shield or two shields, with 25-45 pairs of setae (more than two pairs of posteromedian setae; Fig. 2.13.2A). The sternal shield usually bears two or three pairs of setae; the fourth pair of sternal setae are sometimes on a pair of metasternal shields. The genital shield is usually trapezoidal to subrectangular (Fig. 2.13.1C). Genital opening in males is presternal at the base of tritosternum. The ventrianal shield is subcordate to broadly covering venter. The legs are long, terminating in a pair of claws, which may be absent in leg I.

Several species of Ascidae show promise as biological control agents of soil inhabiting pests in greenhouses. *Lasioseius athiasae* Nawar & Nasr and *Protogamasellus dioscorus* Manson are effective predators of the root-knot nematode (*Meloidogyne javanica*) on kidney bean (*Phaseolus vulgaris*). When released four days before the inoculation of the nematodes onto the plant, they can significantly reduce nematode galls, immature females and the total number of nematodes. *Lasioseius bispinosus* Evans feeds on a variety of small soil animals. It is relatively small compared to *Hypoaspis* and is not as effective a predator against *R. robini*, but can successfully attack the juvenile stages of *R. robini* hidden inside the lily bulb. *Lasioseius fimetorum* Karg feeds on *Tyrophagus putrescentiae*, larvae and pupae of thrips (*Frankliniella occidentalis*), eggs of the Collembola and sciarid larvae (*Bradysia paupera*). When fed on *Tyrophagus putrescentiae* at 20°C, females live for 38.6 days and produce a total of 19.4 eggs.

References & suggested further reading. Lesna *et al.* (1995); Halliday *et al.* (1998); Amin *et al.* (1999); Enkegaard and Brodsgaard (2000).

12.3 Prostigmata

12.3.1 Stigmaeidae

The Stigmaeidae is a family of the superfamily Raphignathoidea. This is a cosmopolitan family and consists of nearly 400 species in 25 genera.

The stigmaeids are small to medium-sized mites, with most species measuring 200-500 μm. They are ovoid or round in shape, and yellow, orange or red in colour. The chelicerae are separate or fused together, with styliform movable digits which is not recurved basally as in spider mites. The stigmata are located at the base of chelicerae but peritremes are absent. The palps are five-segmented; the palpal tibia bears a strong tibial claw. The degree and extent of sclerotization of shields on dorsal idiosoma vary greatly and have been used in generic classification. The prodorsum has three or four pairs of setae, but no trichobothria; a pair of eyes and postocular bodies are present in some species. The dorsal hysterosoma bears five rows of up to 22 setae. Ventral opisthosoma bears one to five pairs of aggenital setae. The genital and anal openings are fused or adjacent, bearing one to three pairs of genital setae and three pairs of pseudanal setae. The legs are five-segmented, terminating in a pair of true claws and an empodium with paired tenent hairs arising from a median shaft. Adult males have slightly tapered idiosoma and an aedeagus as in spider mites.

The life cycle consists of the egg, larva, protonymph, deutonymph and adult stages. Sperm transfer is by copulation as in spider mites. Sex determination is arrhenotoky; unfertilized eggs give rise to males only. Long distance dispersal is by wind.

Mites of the genera *Agistemus* and *Zetzellia* are commonly found on the foliage of higher plants. *Agistemus exsertus* Gonzalez has been demonstrated effective in the control of *Panonychus citri* in a greenhouse release study. It is also a significant predator of *Brevipalus obovatus* and *Aculops lycopersici*. An adult can consume 45 immatures of *A. lycopersici* per day.

References & suggested further reading. Kethley (1982); Osman and Zaki (1986); Yue and Tsai (1995); Rezk and Gadelhak (1996).

12.3.2 Anystidae

The Anystidae is a family of the Anytoidea. It is a cosmopolitan family of generalist predators found on a variety of habitats.

The anystids are medium to large, red or orange mites, with radiating long legs. The chelicerae are independent and each bears two setae; the movable digits are strongly developed and hooked, whereas the fixed digits are completely reduced. The palps are five-segmented with one to three spines distally on the inner face of the tibia. The stigmata are located near the cheliceral bases, with short emergent peritremes. The prodorsum has a naso anteriorly and two pairs of trichobothria and two pairs of normal setae on the prodorsal shield. There are one or two pairs of eyes on the prodorsum. The hysterosoma is often hypertrichous posteriorly. The genital and anal valves are separate in both sexes. Legs terminate in a pair of true claws and a claw-like or cup-like empodium.

The life cycle consists of the egg, larva, protonymph, deutonymph, tritonymph and adult stages. Some species have mobile prelarvae. Sperm transfer is by deposited spermatophores.

A common species, *Anystis baccarum* (L.), is often found on dry surfaces of the ground and plants, feeding on small insects and mites. It also occasionally moves into greenhouses and houses, running on walls and tables. *A. baccarum* has been tested as biological control agents of spider mites. It feeds on all stages of *Tetranychus turkestani*. *Anystis agilis* (Banks) feeds on *Tetranychus urticae*. *Anystis salicinus* L. is an effective predator of *Halotydeus destructor*. The slow developmental rates of anystids makes them unfavourable agents for mass-rearing and releases.

References & suggested further reading. Kethley (1982); Sorensen *et al.* (1976); Khanjani *et al.* (1999); Wallace (1981).

12.3.3 Cunaxidae

The Cunaxidae is a family of the Bdelloidea. They are cosmopolitan and occur in the soil, leaf litter, compost, moss, plants and stored products.

Cunaxids are small to medium-sized mites and are often red or brown in colour. The chelicerae are independent and elongated, each bearing one seta. The movable digits are short and hooked, whereas the fixed digits are reduced. The palps are three to five segmented, raptorial, often armed with long spines on the internal margin. The infracapitulum is elongated with a neck-like constriction laterally. The stigmata is located at the bases of chelicerae without peritremes. The prodorsum has two pairs of prominent trichobothria and one or two pairs of ordinary setae. The genital pore is terminoventral in both sexes with a maximum of two pairs of genital acetabulae. Legs are five-segmented, terminating in a pair of true claws and a rayed empodium. A trichobothrium is present on tibia IV.

The life cycle consists of the egg, larva, protonymph, deutonymph, tritonymph and adult stages. Adults build silk webbing for eggs and their development.

Cunaxids are generalist predators of small arthropods and nematodes. *Coleoscirus simplex* (Ewing) feeds on immature stages and adults of root-knot nematodes (*Meloidogyne*) and small arthropods in greenhouse pot cultures. A generation is completed two weeks at 28°C and adult females produce an average of 4.4 eggs per day. Cannibalism is common, including attacks on quiescent immatures in the moulting webbing. Another species, *Cunaxa setirostris* (Hermann), feeds on active stages of *T. ludeni*. One female predator consumes an average of 330 mobile prey during its life span. This species has good potential as biological control agent of *T. ludeni*.

References & suggested further reading. Kethley (1982); Walter and Kaplan (1991); Arbabi and Singh (2000).

12.3.4 Erythraeidae

The Erythraeidae is a cosmopolitan family of the superfamily Erythraeioidea. The larvae are parasites of arthropods but deutonymphs and adults are free-living predators of small arthropods.

Erythaeids are medium to very large mites, with adults of most species measuring 1,000-3,000 μm. Most species are orange, red and brown in colour and the body is covered with a coat of setae, giving a velvety appearance. The characteristic chelicerae are elongate, styliform and retractile. The palps have a strong tibial claw. A pair of stigmata open near the base of the cheliceral body. The prodorsum bears two pairs of trichobothria in two sensory areas connected by a strongly sclerotized crista metopica. Lateral to the crista, there are one or two pairs of sessile eyes. Legs are long with a pair of true claws. Larvae are heteromorphic. They are often red in colour when alive and 300-500 μm long when unfed. The chelicerae are not elongated; the movable digits are short and the fixed digits are reduced. A prodorsal shield is present on the prodorsum, bearing two pairs of trichobothria and two to several pairs of normal setae. One or two pairs of eyes are located lateral to the prodorsal shield. The coxae of leg I and leg II are well separated. There is no anal opening on the ventral opisthosoma. The legs are very long and terminate in a pair of lateral claws and a median claw-like empodium.

The life cycle consists of the egg, larva, protonymph, deutonymph, tritonymph and adult stages. The protonymphs and tritonymphs are non-feeding, quiescent stages. There are often only one or two generations per year. Sperm transfer is by deposited spermatophores.

Deutonymphs and adults are predatory and are common in soil and litter. Species of *Balaustium* are common in orchards and often move onto plants and can feed on pollen. They also occasionally invade greenhouses and feed on small insects, mites and pollen. They may even attack humans.

At 25°C in the laboratory, *Balaustium putmani* Smiley can develop from egg to adult in 39 days. Females each lay 175 or more eggs. In the laboratory, a female predator can consume over 100 eggs and 25 adults of *Panonychus ulmi*. This species can also feed on eriophyid mites and other spider mites such as *Bryobia* and *Tetranychus*.

References & suggested further reading. Newell (1963); Putman (1970); Codagan and Laing (1977, 1981); Childers and Rock (1981).

12.3.5 Tydeidae

This family is introduced in Chapter 9 because some members of this family have been recorded as causing significant damage by feeding on leaves. However, some members of this family are also predators of pest mites. *Homeopronematus anconai* (Baker) attacks *Aculops lycopersici* (Massee), a pest of tomato in greenhouses (Chapter 7), and can significantly reduce its density. *Pronematus ubiquitus* (McGregor) is known to be associated with *A. lycopersici* but nothing is known about its effectiveness as a predator of *A. lycopersici*.

References & suggested further reading. Hessein and Perring (1986).

12.4 Reference cited

Al-Amidi, A.H.K. and Downes, M.J. (1990) *Parasitus bituberosus* (Acari: Parasitidae), a possible agent for biological control of *Heteropeza pygmaea* (Diptera: Cecidomyiidae) in mushroom compost. *Experimental & Applied Acarology* 8, 13-25.

Al-Amidi, A., Dunne, R and Downes, M.J. (1991) *Parasitus bituberosus* (Acari: Parasitidae): an agent for control of *Lycoriella solani* (Diptera: Sciaridae) in mushroom crops. *Experimental & Applied Acarology* 11, 159-166.

Amin, A.W., Mowafe, M.H. and Fatma, S.A. (1999) Effect of predaceous mesostigmatid mites in the control of *Meloidogyne javanica* root-knot nematode on kidney bean. *Pakistan Journal of Nematology* 17, 91-96.

Arbabi, M. and Singh, J. (2000) Studies on biological aspects of predaceous mite *Cunaxa setirostris* on *Tetranychus ludeni* at laboratory condition in Varanasi, India. *Journal of Agriculture and Rural Development* 2(1-2), 13-23.

Cadogan, B.L. and Laing, J.E. (1977) A technique for rearing the predaceous mite *Balaustium putmani* (Acari: Erythraeidae), with notes on its biology and life history. *The Canadian Entomologist* 109, 1535-1544.

Cadogan, B.L. and Laing, J.E. (1981) A study of *Balaustium putmani* (Acari: Erythraeidae) in apple orchards in southern Ontario. *Proceedings of the Entomological Society of Ontario* 112, 13-22.

Childers, C.C. and Rock, G.C. (1981) Observations on the ocurrence and feeding habits of *Blaustium putmani* (Acari: Erythraeidae) in North Carolina apple orchards. *International Journal of Acarology* 7, 63-68.

Enkegaard, A. and Brodsgaard, H.F. (2000) *Lasioseius fimetorum*: a soil-dwelling predator of glasshouse pests? *BioControl* 45, 285-293.

Evans, G.O. (1992) *Principles of Acarology.* CAB International, Wallingford, UK. 563 pp.

Halliday, R.B., Walter, D.E. and Lindquist, E.E. (1998) Revision of the Australian Ascidae (Acari: Mesostigmata). *Invertebrate Taxonomy* 12, 1-54.

Hessein, N.A. and Perring, T.M. (1986) Feeding habits of the Tydeidae with evidence of *Homeopronematus anconai* (Acari: Tydeidae) predation on *Aculops lycopersici* (Acari: Eriophyidae). *International Journal of Acarology* 12, 215-221.

Karg, W. (1961) Ökologische untersuchungen von edaphischen Gamasiden (Acarina, Parasitiformes). *Pedobiologia* 1, 77-89.

Kethley, J. (1982) Acariformes. In: Parker, S.P. (ed.) *Synopsis and Classification of Living Organisms.* McGraw-Hill, New York, pp. 124-125.

Khanjani, J., Kamali, K. and Sahragard, A (1999) Functional response of *Anystis baccarum* (L.) (Acari: Anystidae) to different densities of two spotted spider mite, *Tetranychus turkestani* U. & N. (Acari: Tetranychidae). *Agricultural Sciences and Technology* 13, 141-147.

Lesna, I., Sabelis, M.W., Bolland, H.R. and Conijn, C.G.M. (1995) Candidate natural enemies for control of *Rhizoglyphus robini* Claparede (Acari: Astigmata) in lily bulbs: exploration in the field and pre-selection in the laboratory. *Experimental & Applied Acarology* 19, 655-669.

Newell, I.M. (1963) Feeding habits in the genus *Balaustium* (Acari: Erythraeidae), with special reference to attacks on man. *Journal of Parasitology* 49, 498-502.

Osman, A.A. and Zaki, A.M. (1986) Studies on the predation efficiency of *Agistemus exsertus* Gonzalez (Acarina, Stigmaeidae) on the eriophyid mite, *Aculops lycopersici* (Massee). *Anzeiger für Schädlingskunde, Pflanzenschutz, Umweltschutz* 59, 135-136.

Putman, W.L. (1970) Life history and behavior of *Balaustium putmani* (Acarina: Erythraeidae). *Annals of the Entomological Society of America* 63, 76-81.

Rezk, H.A. and Gadelhak, G.G. (1996) Relationship between phytophagous and predatory mites in citrus orchards and the effect of acaricides on their populations. *Alexandria Journal of Agricultural Research* 41, 217-224.

Sorensen, J.T., Kinn, D.N., Doutt, R.L. and Cate, J.R. (1976) Biology of the mite *Anytis agilis* (Acari: Anystidae): a California vineyard predator. *Annals of the Entomological Society of America* 69, 905-910.

Walter, D.E. and Kaplan, D.T. (1991) Observations on *Coleoscirus simplex* (Acarina: Prostigmata), a predatory mite that colonizes glasshouse cultures of rootknot nematode (Meloidogyne spp.), and a review of feeding behavior in the Cunaxidae. *Experimental & Applied Acarology* 12, 47-59.

Yue, B.-S. and Tsai, J.-H. (1995) *Agistemus exsertus* Gonzalez (Acari: Stigmaeidae) as a predator of citrus red mite (*Panonychus citri* [McGregor]). *Journal of the New York Entomological Society* 103, 107-113.

Part IV
Accessories

- Glossary of terms used in the book
- Mite information sources
- Index

Glossary 13

Accetabula: Rounded sclerites, in pairs, such as those alongside of genital openings.

Adanal segment (AD): Segment XIV of acariform mites added in the protonymph, bearing the fourth pair of legs.

Adoral setae: Setae near the mouth on the lateral lips.

Adult: Sexually mature individual; the last stage of the life cycle.

Aedeagus: Penis, the male genital organ used in copulation.

Aggenital: Of the region on two sides of the genital region.

Ambulacrum (ambulacra *pl.***)**: The ambulatory appendage at the distal end of the tarsus, usually consists of a pair of lateral claws and an empodium.

Anal segment: Segment XV of acariform mites added in the deutonymph.

Anal suckers: Rounded sclerites, in pairs, on two sides of the anus.

Anamorphosis (anamorphic *adj.***)**: Type of metamorphosis in which segments are added posteriorly in the course of postembryonic development.

Anus (anal *adj.***)**: External opening of the hindgut for excretion.

Apodeme: Internal sclerotized projection of the exoskeleton, often serving as attachment sites for muscles (e.g. apodemes in coxisternal regions of Tarsonemidae and Acaridae).

Apotele: Terminal segment of appendages attached to the tarsus.

Arrhenotoky (arrhenotokous *adj.***)**: Form of parthenogenesis in which unfertilized eggs give rise to males.

Basifemur: Proximal part of the divided femur.

Bothridium: Cuticular structure in the form of a cavity where bothridial seta (sensillus) is inserted.

Brooming: The formation of foreshortened terminal branches on a plant, with many growing points and branches.

Bursa copulatrix: Genital pouch in females of Astigmata with a copulatory opening for accepting sperm (its opening is not the genital opening).

c setae: The first row of setae on the dorsal opisthosoma of a typical acariform mite. Three pairs (c_1, c_2 and c_3) are usually present.

Capitulum: The anteriormost or the "head" part of the body, i.e. the gnathosoma. Mites lack a true head.

Chaetotaxy (chaetotactic *adj.***)**: The number and pattern of distribution of setae.

Chelate-dentate: Pincer-like and with teeth.

Chelicera (cheliceral *adj.***)**: The anterior pair of appendages of the gnathosoma that are used to pierce or chew prey.

Cheliceral body: The principal (the second) segment of chelicera.

Cephalothorax: The fused head and thorax as in a spider.

Claw: The lateral pair attached to the pretarsi of legs.

Condylophore: A pair of stout parallel sclerites, connecting the empodial claw with the tarsal base.

Corniculus (corniculi *pl.***)**: Enlarged, often horn-shaped seta on the apical part of the infracapitulum.

Coxa (coxal *adj.***)**: The basal segment of legs and palps.

Crista metopica: Strongly sclerotized rod-like structure in the middle of the prodorsum in many Trombidiodea and Erythraeidae.

Cupule: See lyrifissure.

Cuticle (cuticular *adj.***)**: Outmost layer (exoskeleton) covering the body of mites.

d setae: The second row of setae on the dorsal opisthosoma of a typical acariform mite. Two pairs (d_1 and d_2) are usually present.

Deuterogyny (deuterogyne *n. & adj.***)**: The presence of two forms of females in some eriophyoid mites in which the second type (deuterogyne, the overwinter or oversummer type) is morphologically different from normal females, has no male equivalent and can resist extreme temperatures.

Deutonymph: The second nymph (see hypopus).

Deutosternum (deutosternal *adj.***)**: A groove or gutter-like structure on the subcapitulum of parasitiform mites, often with transverse rows of anteriorly-directed denticles.

Diapause: A period of greatly decreased metabolic activity with arrestment of development and reproduction, often induced by short day length and low temperature.

Dimorphism: Two different forms (e.g. sexual dimorphism for males and females with different structures).

Diplo-diploidy: Form of bisexual reproduction in which both males and females are diploid (having two sets of chromosomes).

Diploid: Having two sets of chromosomes.

Disjugal furrow: The groove separating the prosoma and opisthosoma.

Dorsum: The dorsal surface of the body or appendages.

Duplex setae: The pair of setae found on tarsus I and tarsus II, often with joint bases; one of the pair is usually very long (solenidion) and the other a small tactile seta.

e setae: The third row of setae on dorsal opisthosoma of a typical acariform mite. Two pairs (e_1 and e_2) are usually present in spider mites.

Empodium (empodial *adj.*, **empodia** *pl.*): The median structure attached to the pretarsi of legs, often claw-like or pad-like when present.

Emulsifiable concentrate: The formulation of pesticides applied diluted in water.

Epigynial shield: Genital shield.

Epizootics: Large-scale outbreaks of diseases.

Eupathidium (eupathidia *pl.*, **eupathidial** *adj.*): A hollow smooth seta found on leg and palp tarsi.

f setae: The fourth row of setae on the dorsal opisthosoma of a typical acariform mite. Two pairs (f_1 and f_2) are usually present.

Famulus (famuli *pl.*): A small, hollow seta on tarsus I and tarsus I of acariform mites.

Femur (femora *pl.*): The fourth segment of legs and palps counting from the distal end in a typical mite; it is divided into a distal telofemur and proximal basifemur in some groups.

Femorogenu (femorogenua *pl.*, **femorogenual** *adj.*): Union of femur and genu.

Fixed digit: Non-movable distal part of the principal segment of the chelicera.

Flange: Posterolateral outgrowth of femorogenu IV in some tarsonemid males.

Functional response: Changes in the rate of prey attacked per predator in relation to prey density.

Furrow: Groove separating parts of the body or segments of the appendages.

Genu (genua *pl.*): The third segment of legs and palpi counting from the distal end in a typical mite, distal to femur and proximal to tibia.

Gnathosoma: The part of the body anterior to the idiosoma, bearing the palps and chelicerae used in feeding.

Grandjean's organ: Sclerotized structure above trochanter I in some Astigmata, often extending anterior into various shapes.

h setae: The fifth row of setae on dorsal opisthosoma of a typical acariform mite. Two to three pairs (h_1-h_3) are usually present.

Haller's organ: A sensory organ for chemoreception on the distodorsum of tarsus I in ticks.

Haplo-diploidy: Form of bisexual reproduction in which males are haploid (having one set of chromosomes) and females are diploid (having two sets of chromosomes).

Heteromorphic: Having different forms in the same species (e.g. larva versus adult in Erythraeidae, hypopus versus adults in Acaridae).

Holodorsal shield: A large dorsal shield covering both the anterior and posterior parts of the idiosoma.

Homeomorphic: Having the same form.

Hypopus (hypopi *pl.*, **hypopal** *adj.*): Non-feeding deutonymph in the Astigmata adapted for dispersal and resisting adverse environmental conditions.

Hysterosoma: The part of the body posterior to the sejugal furrow.

Idiosoma: Main part of the body posterior to the gnathosoma.

Infracapitulum: The inferior part of the gnathosoma, bearing the lips and palps, and containing the mouth and pharynx.

Intrinsic rate of increase (r_m): The rate of populaton increase under optimal conditions; a function of age-specific survival and reproductive rates.

Kairomone: Chemical emitted by an organism as a specific signal and being beneficial to another organism of another species.

Lacinia (laceniae *pl.*): Seta-like structure, often in pairs, arising from the base of the tritosternum in parasitiform mites.

Larva (larvae *pl.*, **larval** *adj.*): Immature stage between the egg and the nymphal stages, often bearing three pairs of legs.

Lobe: Projections of various shapes from integument striae.

Lyrifissure: Slit sensory organ in the cuticle of the idiosoma and appendages. Also known as cupule.

Macroseta: Enlarged setae on genu, tibia and basal tarsus of legs in Mesostigmata.

Mesonotal scutellum (mesonotal scutellae *pl.*): Platelets, often in pairs, in the middle of the idiosomal dorsum.

Metapodosoma (metapodosomal *adj.*): Part of podosoma posterior to the sejugal furrow, bearing legs III and IV.

Metasternal shield: Small plate posterior to sternal shield, often bearing the fourth pair of sternal setae in Mesostigmata.

Microseta: A small seta often located at the distal part of tibia I, genu I and genu II of some acariform mites.

Microtubercle: Cuticular lobe on annulated striae in some eriophyoid mites.

Mouth parts: Structures in distal gnathosoma involved in food-intake.

Movable digit: The inferior segment (apotele) of the chelicera that is movable and basally hinged to the fixed digit to form the pincer.

Naso: Unpaired protuberance on distal prodorsum in some acariform mites. Also as nasus.

Notation: Methods of denoting structures using a system of letters, abbreviations, numbers and symbols.

Notogaster: Dorsal plate on the hysterosoma of oribatid mites.

Nymph: Immatural stage(s) between the larva and adult in a life cycle.

Opithosoma (opisthomatic *adj.***)**: Part of the body posterior to the podosoma.

Opisthonotal shield: Dorsal plate on the opisthosoma.

Oviposition: Deposition of eggs.

Palp: The second pair of appendages of the gnathosoma, used in sensing and handling food items. Also as pulpus.

Parthenogenesis (parthenogenetic *adj.***)**: The form of reproduction without fertilization of the egg.

Peranal segment (PA): Segment XVI added in tritonymph in acariform mites.

Peritreme (peritrematic *adj.***):** Gutter-like or tube-like structure associated with a stigmata.

Pharynx (pharyngeal *adj.***)**: Suctorial portion of the alimentary canal with muscular walls, extending from the mouth to the oesophagus.

Pheromone: Chemical emitted by an organism as a specific signal to another organism of the same species.

Phoresy (phoretic *adj.***)**: Transport of one organism by another without parasitism.

Phoretomorph: Forms adapted for phoresy.

Physogastry (physogastric *adj.***)**: Excessive enlargement of the opisthosoma in some gravid tarsonemid mites.

Pilus dentilis: Seta on the fixed digit of the chelicera.

Podonotal shield: Dorsal plate on the podosoma.

Podosoma: The part of the idiosoma bearing the legs.

Podospermy: Form of copulation in which sperm is transferred by the spermatodactyl to openings on coxae III of the female.

Precopulation: Behaviour of males of some tarsonemids in finding pharate females still contained in the larval cuticle and carrying them before copulation.

Pregenital: Of the region anterior to the genital pore, belonging to segment VII of an acariform mite.

Pretarsus: Less sclerotized, distal part of the tarsus, which forms part of the ambulacrum, and contains an endoskeleton (often a pair of sclerites) for articulating with the apotele.

Prodorsum: The dorsum of the idiosoma anterior to the disjugal furrow.

Propodosoma: The anterior part of the idiosoma bearing legs I and II.

Prosoma: The part of the body anterior to the opisthosoma, including the gnathosoma and podosoma.

Protonymph: The first nymph.

ps setae: Pseudanal setae, the sixth row of setae on the dorsal opisthosoma of a typical acariform mite. Two or three pairs (ps_1, ps_2 and ps_3) are usually present. The last two pairs are often ventral in location.

Pseudanal segment (PS): Segment XIII of an acariform mite, bearing up to three pairs of setae.

Pseudanal setae: See ps setae.

Pseudo-arrhenotoky: Form of sexual reproduction in which males are from fertilized eggs, but one set of chromosomes is deleted later.

Pteromorph: Wing-like or ear-like dorsolateral extension of the notogaster in some oribatid mites.

Pulvilus (pulvili *pl.*, **pulvilar** *adj.***)**: Cushion-like structure located ventral of and between the paired claw, and forming part of the ambulacrum.

Pygidial shield: Doral plate on the posterior part of the opisthsoma.

Receptaculum seminis: Organ in females for receiving the sperm from males.

Rutellum (rutella *pl.***)**: Enlarged seta inserted on the laterodistal infracapitulum in some acariform mites.

sc setae: The second (scapular) row of setae on the prodorsum of a typical acariform mite. Two pairs (sc_1 and sc_2) are present.

Sclerite (sclerotized *adj.*, **sclerotization** *n.***)**: Part of hardened integument; plate; shield.

Segment: Each of the successive elements of the body or appendages.

Sejugal furrow: The furrow separating the propodosoma and hysterosoma.

Sex ratio: Ratio between males and females, expressed in various ways (e.g. proportion of female offspring of a mother).

Solenidion (solenidia *pl.***)**: A hollow sensillary seta with thin wall, often smooth externally and annulated internally, on the tarsus and tibia and genu of legs and also on the palpal tarsus.

Sensillus (sensilli *pl.***)**: Bothridial seta. Also as sensillum (sensilla *pl.*).
Spermatodactyl: The finger-like structure on the movable digit in males of Demanyssina (Mesostigmata) for introducing sperm into females.
Spermatophore: Capsule containing the sperm.
Spermatotreme: The finger-like structure on the movable digits in males of Parasitina (Mesostigmata) for introducing sperm into females.
Spinneret: Enlarged eupathidium on distal palptarsus and used in spinning webbing by some species of spider mites.
Spirotheca (spirothecae *pl.***)**: Pouch-like structure for carrying fungal spores in fungivorous mites (e.g. Pygmephoridea).
Stage: Distinct period separating the successive moults in the life cycle of mites (e.g. egg stage, larval stage, nymphal stage and adult stage).
Stigmata (stigmatic *adj.***)**: External opening of the respiratory system.
Stylet: Needle-like movable digit of the chelicera that is specialized for piercing.
Subcapitulum (subcapitular *adj.***)**: The ventral surface of infracapitulum.
Supracoxal seta: Setae associated with the base of the palp, leg I and leg II in acariform mites.
Tactile: Pertaining to the sense of touch.
Tarsus (tarsi *pl.***)**: The subterminal segment of legs and palps, distal to the tibia and bearing the apotele.
Tegula: Often tongue-shaped, posteromedian extension of the consolidated coxisternal plate IV in Tarsonemidae.
Telofemur: Distal part of the divided femur.
Tenent hairs: Slender hairs arising from claws or empodia, thought to allow mites to hold on to leaf surface; their distal end often slightly enlarged in many species of spider mites.
Thelytoky: Form of parthenogenesis in which females produce females without males.
Thumb-claw complex: Structure at the distal end of the palp, formed by the translocation of the tarsus to the basal part of the tibia. It is used to hold food items. Also as thumb-claw process.
Tibia (tibiae *pl.***)**: The second segment of legs and palps counting from the distal end in a typical mite. It is joined distally with the tarsus and basally with the genu.
Tibiotarsus: Union of the tarsus and tibia.
Trichobothrium (trichobothria *pl.***)**: A compound structure consisting of a cavity (bothrium) and a seta arising from it (bothridial setae).
Tritonymph: The third nymph.
Tritosternum: Structure arising from base of the infracapitulum, typically with a slender base branching into a pair of pilose lancinae.

Trochanter (trochanteric *adj.*): The fifth segment of legs and palps counting from the distal end in a typical mite. It is joined distally with the femur and basally with the coxa.

Tocospermy: Form of copulation in which sperm is transferred directly from the male genital orifice to that of the female.

Tolerance: A natural lack of susceptibility to pesticides, diseases or pests.

v setae: The first (vertical) row of setae on the prodorsum of an acariform mite. Two pairs (v_1 or *sci* and v_2 or *sce*) are usually present.

Ventrianal shield: Union of ventral shield and anal shield in Mesostigmata.

Wettable powder: The formulation of pesticides applied as power suspended in water.

Whirl: A cycle of dorsal, lateral and ventral setae on leg segments.

Appendix 14

Mite Information Sources

14.1 Introduction

This chapter provides directions for readers who are interested in finding more information about acarological societies, journals, directories, websites, collections, courses and suppliers of mite predators for biocontrol.

14.2 Professional societies/organizations

14.2.1 International Congress of Acarology (ICA)

ICA is held every four years in countries selected by the Executive Committee of ICA. Proceedings are published after each Congress (see 14.3.1). The ICA also compiles a directory of acarologists in the world (see 14.4.1).

14.2.2 Systematic and Applied Acarology Society (SAAS)

An international society dedicated to promoting the development of acarology and facilitating collaboration and exchange of information among acarologists in different parts of the world. SAAS publishes the journal *Systematic and Applied Acarology, Systematic and Applied Acarology Special Publications* and the newsletter *Acarology Bulletin*.

www.nhm.ac.uk/hosted_sites/acarology/saas/

14.2.3 Acarological Society of America (ASA)

An international society devoted to furthering all aspects of the knowledge of mites. ASA publishes an irregular newsletter and organizes an annual meeting with the Entomological Society of America.

www.wm.edu/biology/mites/

14.2.4 European Association of Acarologists (EURAAC)

An international organization for acarologists in Europe to provide a means for personal contact among acarologists, to provide support for instructional workshops for those wishing to specialize in acarology, and to represent the views of European acarologists in approaches to official bodies in matters pertaining to the interests of acarology. EURAAC publishes an irregular newsletter and organizes symposia in Europe at regular intervals for the dissemination of knowledge among acarologists. Proceedings of the symposia were published as a series of books.

www.fu-berlin.de/euraac/

14.2.5 Société des Acarologues de Langue Francaise (SIALF)

An international society to promote research in acarology by exchanges of information among French-speaking members, to support growth of scientists and development of acarology, and to initialize publications of applied or fundamental research. SIALF organizes the 'International Courses of Acarology' regularly and publishes an occasional journal newsletter, *Salfia*.

www.nhm.ac.uk/hosted_sites/acarology/saas/Hosted/sialf/index.htm

14.2.6 Sociedad Latinoamericana de Acarologia (SLA)

An international society for better communication and development of acarology in Latin America. SLA organizes regular meetings for members.

www.slacarologia.com

14.2.7 African Acarology Association (AAA)

An international society with the mission to provide a means of communication, cooperation and personal contact amongst acarologists in Africa, to promote and support the study of mites and ticks in Africa, and to provide a formal platform or forum for reports and discussion of research activities, ideas and methods, research aims and needs. AAA organizes symposia in Africa at regular intervals.

www.nhm.ac.uk/hosted_sites/acarology/saas/Hosted/aaa/

14.2.8 Regional acarological organizations

Several countires have their own acarological societies. Some are quite active. The Acarological Society of Japan holds annual meetings and symposia, and publishes *Journal of Acarological Society of Japan*. The Acarological Society of India publishes *Journal of Acarology* and *Acarology Newsletter* and organizes symposia irregularly.

14.3 Serial publications

14.3.1 Proceedings of the International Congress of Acarology

Eleven congresses have been held so far and proceedings of ten of these have been published (in order of date of publication):

Wooley, T.A. (ed.) (1964) *Proceedings [of] 1st International Congress of Acarology*, Fort Collins, Colorado, USA, 2-7 September 1963. *Acarologia* 6, 1-439.
Evans, G.O. (ed.) (1969) *Proceedings of the 2nd International Congress of Acarology*, Sutton Bonington, England, 19-25 July, 1967. Akademiai Kiado, Budapest. 652 pp.
Anonymous (1969) *Contributions to the Symposia of the 2nd International Congress of Acarology*, Sutton Bonington, 1967. *Acarologia* 11, 355-503.
Daniel, M. and Rosicky, B. (eds) (1973) *Proceedings of the 3rd International Congress of Acarology held in Prague, August 31 - September 6, 1971*. Academia, Prague. 837 pp.
Piefl, E. (ed.) (1979) *Proceedings of the 4th International Congress of Acarology, Saalfelden (Austria)* [12-19 August 1974]. Akademai Kiado, Budapest. 752 pp.
Rodriguez, J.G. (ed.) (1979) *Recent Advances in Acarology*. Academic Press, New York, Vol. 1, 631 pp., Vol. 2, 569 pp. [A selection of papers read at the 5th International Congress of Acarology, East Lansing, Michigan, USA, August 1978].
Griffiths, D.A. and Bowman, C.E. (eds) (1984) *Acarology VI*. Ellis Horwood Limited, Chichester, Vol. 1, 645 pp., Vol. 2, 650 pp. [Proceedings of the 6th International Congress of Acarology, Edinburgh, Scotland, 5-11 September 1982].
Channabasavanna, G.P. and Viraktamath, C.A. (eds) (1989) *Progress in Acarology* (Proceedings of the VII International Congress of Acarology held in 1986 in India). E.J. Brill, Leiden, Vol. 1 484 pp., Vol. II 532 pp.
Dusbabek, F. and Bukva, V. (eds) (1991) *Modern Acarology I & II* (Proceedings of the VIII International Congress of Acarology, Ceske Budejovece, Czechoslovakia, August 6-11, 1990). Academia, Prague, Vol. I 651 pp., Vol. II 779 pp.
Needham, G.R., Mitchell, R., Horn, D.J. and Welbourn, W.C. (eds) (1999) *Acarology IX: Symposia*. Ohio Biological Survey, Columbus, Ohio. 507 pp.
Mitchell, R., Horn, D.J., Needham, G.R. and Welbourn, W.C. (eds) (1999) *Acarology IX: Proceedings*. Ohio Biological Survey, Columbus, Ohio. 718 pp.
Halliday, R.B., Walter, D.E., Proctor, H.C., Norton, R.A. and Colloff, M.J. (eds) (2000) *Acarology. Proceedings of the 10th International Congress*. CSIRO Publishing, Melbourne. 657 pp.

14.3.2 *Acarologia*

A quarterly international journal published in France, with papers on all aspects of mites and ticks in English, French and German.

http://alor.univ-montp3.fr/acrlg/Acrlg024/Acrlg.htm

14.3.3 *International Journal of Acarology*

A quarterly international journal published in the USA, with papers on all aspects of mites and ticks in English.

http://pw2.netcom.com/~v.prasad/jinfo.html

14.3.4 *Experimental and Applied Acarology*

A monthly international journal published in English in The Netherlands, with papers on experimental and applied aspects of the Acari.

www.kluweronline.com/issn/0168-8162

14.3.5 *Systematic and Applied Acarology*

An annual international journal published in the UK, with papers on systematic and applied aspects of mites and ticks in English.

www.nhm.ac.uk/hosted_sites/acarology/saas/saa.html

14.3.6 *Systematic and Applied Acarology Special Publications*

A rapid international journal published in the UK, with papers on systematic and applied aspects of mites and ticks in English. All papers are free online from the journal website.

www.nhm.ac.uk/hosted_sites/acarology/saas/saasp.html

14.3.7 *Acarina: Russian Journal of Acarology*

A biannual journal published in Russia with papers on all aspects of mites and ticks in English and Russian.

www.orc.ru/~kmkweb/acarina.htm

14.3.8 *Journal of the Acarological Society of Japan*

A biannual journal published in Japan with papers on all aspects of mites and ticks in English and Japanese. Formerly *Proceedings of the Japanese Association for Acarology.*

www.affrc.go.jp:8001/acari/journal/journal.html

14.4 Directories of acarologists

14.4.1 ICA Directory of Acarologists of the World 2002

A list of 2,531 acarologists compiled by Valerie Behan-Pelletier for the XI International Congress of Acarology in 2002. It is online.

www.nhm.ac.uk/hosted_sites/acarology/ica/directory/index.html

14.4.2 Online Register and Directory of Acarologists

An online entry form for a database of acarologists and online directory of acarologists in the world by Zhi-Qiang Zhang.

www.nhm.ac.uk/hosted_sites/acarology/database/index.html
http://internt.nhm.ac.uk/cgi-bin/acar/list.dsml?lastname=A%25

14.5 Internet resources

14.5.1 Web resources on mites of greenhouses

A special website by Zhi-Qiang Zhang containing reference material for this book, with links to websites, colour photographs and others. The website will be regularly updated.

www.nhm.ac.uk/hosted_sites/acarology/zhang/greenhousemites/

14.5.2 E-mail acarology discussion list

Named ACAROLOGY, this discussion list is dedicated to promoting the information exchange and communication among all those who are interested in the Acari. It is owned and managed by Zhi-Qiang Zhang and all messages on the list are achieved and available online.

www.nhm.ac.uk/hosted_sites/acarology/acarolist.html

14.5.3 The acarology home page

A general site for acarology developed by Zhi-Qiang Zhang with various kinds of information on mites and ticks.

www.nhm.ac.uk/hosted_sites/acarology/

14.6 Acarological collections

Mite specimens are scattered around in many collections and museums in the world. The following is an excellent guide.

Kethley, J. (ed.) (1979) *An Index to the Acarina Collections of the World (to the Family Level)*. NAMRU-3, Cairo. 87 pp.

14.7 Acarology courses

14.7.1 The Ohio State University Acarology Summer Program

Short courses of one week to three weeks in English. Held annually in June-July. Workshops offered include: Introductory acarology (1 week); Agricultural acarology (2 weeks); Soil acarology (3 weeks); Medical-veterinary acarology (2 weeks).

www.biosci.ohio-state.edu/~acarolog/sum2k1.htm

14.7.2 International Courses of Acarology

Organized by SIALF irregularly, often on special topics in French. Contact person: M. Bertrand, Zoogeographie Univ Montpellier 3, BP5043 F-34032, Montpellier, France.

14.8 Suppliers of natural enemies for biological control

Many commercial companies supply predatory mites for biocontrol control of mites and insects, and also other natural enemies for mite control (Anonymous, 2000). Common predatory mites used in greenhouses are available from many suppliers: e.g. *Phytoseiulus persimilis* 49 suppliers, *Neoseiulus cucumeris* 29 suppliers, *Neoseiulus californicus* 23 suppliers and *Hypoaspis miles* 15 suppliers.

Anonymous (2000) 2001 directory of least-toxic pest control products. *IPM Practitioners* 21(11-12), 1-48.

Index

15

Page numbers in bold refer to figures.

Abamectin 109, 113, 118, 119, 132, 151, 185
Acarapinae 99
Acarapis 100
Acarapis woodi 11
Acaridae 141
Acariformes 19
Acarus 141
Aceria 136
Aceria dianthi 136
Aceria georghioui 136
Aceria lycopersici 135
Aceria paradiathi 136
Achaearanea tepidariorum 67
Aculops lycopersici 7, 128, 129, **130**, 213, 216
Aedeagus 48
Africa 68, 90, 110, 118, 119, 193
African violet 104, 110, 118
Ageratum conyzoides 185
Agistemus 213
Agistemus exsertus 73, 92, 132, 213
Aldicarb 185
Aldrin 151
Algeria 68, 180
Amaryllidaceae 115
Amaryllis 115, 116
Amblyseiinae 171, **174**
Amblyseius californicus 190
Amblyseius cucumeris 171
Amblyseius largoensis **177**, 194
Ambulacrum 17
Ambulatory appendage **18**
Amitraz 113, 119
Angelica utilis 66
Angola 90
Anus 17
Anystidae 69, 214
Anystis agilis 214
Anystis baccarum 69, 214
Anytoidea 214
Arachnida 11
Argentina 90, 136
Arrhenotokous 205, 213
Ascidae **28**, 212
Ascoidea 212
Asia 68, 70, 90, 110, 119, 192, 204, 206
Asparagus 6
Asplenium 114
Astigmata 19, **22**, 29, 141
Aubergine 61, 129, 180
Australasia 90, 110, 191, 192
Australia 20, 66, 70, 74, 90, 166
Austria 90, 182
Avermectin 109
Avocado 70
Azadirachta indica 66
Azalea 70, 108, 110, 118, 119
Azinphos-ethyl 151
Azocyclotin 109, 149

Bacillus thuringiensis 132
Balaustium 216
Balaustium putmani 216
Bdelloidea 214
Bean 64, 68, 74, 181, 183
Bean mite 64
Beauveria bassiana 108
Bedding plants 6
Begonia 104, 110, 151
Belgium 118, 183
Bemisia argentifolii 107
Bemisia tabaci 107, 187, 194
Benzoximate 154
Berlese-Tullgren funnel 36
Bifenthrin 119
Binapacryl 132
Biological control 57, 58, 59, 66, 92, 93, 95, 108, 112, 131, 136, 149, 171, 180, 181, 183, 188, 189, 190, 193, 194, 203, 206, 212, 214, 215
Biology 48, 56, 62, 88, 92, 95, 96, 97, 100, 107, 111, 118, 120, 128, 131, 134, 135, 136, 142, 148, 150, 157, 158, 172, 180, 186, 191, 194, 204, 205, 207
Bolivia 69
Bothridium 20
Bradysia 206, 208
Bradysia paupera 213
Brazil 109
Brevipalpinae 87
Brevipalpus 87, **89**, 96
Brevipalpus obovatus **89**, 90, **91**, 213
Brevipalpus phoenicis 89, 93, **94**
Brevipalpus russulus **89**, 96
Broad mite 7, 104
Bromopropylate 109, 132
Brussels sprout 75
Bryobia **51**, 52, 74, 75, 216
Bryobia calida **52**, 75
Bryobia kissophila **52**, 74
Bryobia lagodechiana **52**, 74
Bryobia praetiosa **52**, 75
Bryobinae 47
Bulb mite 149
Bulgaria 68, 129, 182, 183
Byelorussia 118, 119

Cabbage 6
California 136
Campanula 90
Canada 69, 74, 90, 118
Canada Balsam 41
Cannibalism 215
Capitulum 12
Capsicum 7, 108, 109, 151, 184
Carbofuran 151
Carmine spider mite 7, 61
Carnation 7, 61, 110, 136, 164, 185
Carnation bud rot 164
Cassava 109
Celery 6
Central America 68, 119
Cephalothorax 11
Cereals 75, 164, 166
Chaeototaxy 18
Cheletogenes ornatus 92
Chelicerae 14, 47, 211, 212
Chelicerata 11
Chemical control 59
Cheyletidae 92
Chile 69, 70, 74, 180
Chilean predatory mite 180
Chillies 129
China 66, 68, 70, 73, 109, 118, 119, 120, 185, 191
China-berry 73
Chinese cabbage 119
Chinomethionat 109
Chlorfenapyr 118
Chlorobenzilate 75, 135, 136
Chlorpyrifos 118, 151
Chrysanthemum 104, 110, 128, 133, 134, 156, 188
Chrysanthemum leaf rust mite 133
Chrysoperla mutata 69

Chrysoperla sinica 73
Cineraria 154
CIS 68, 74
Cissus 90, 118
Citrus 69, 70, 71, 90
Citrus red mite 71, 73
Citrus silver mite 104
Citrus thrips 193
Clofentezine 59
Clorobenzilate 93
Clover 75
Coccinellids 69
Coleoscirus simplex 215
Collecting mites 33
Collembola 212, 213
Colombia 69, 107
Condylophore **18**
Conifers 73
Consolida ambigua 155
Controlled release system 187
Convolvulus 129
Corniculi 20
Costa Rica 68, 69, 74, 119
Crassula 113
Cryptostigmata 19
Cuba 73, 109
Cucumber 61, 68, 74, 75, 104, 119, 151, 152, 154, 156, 157, 180, 181, 182, 187, 193, 208
Cucurbits 6, 7
Cultural and physical methods 60
Cunaxa setirostris 66, 215
Cunaxidae 66, 214
Cyclamen 104, 110, 113, 118, 152, 156, 207
Cyclamen mite 109, 191
Cyhexatin 132, 149, 153
Cymbidium 152, 155, 156, 185
Cymbidium clauboda 155
Cypermethrin 108
Cyprus 90, 136

Dahlia 104, 147
Dark-red spider mite 64
Datura 129
Delphinium ajacis 155
Deltamethrin 151
Dendrolaelaps sayedi 149
Denmark 118
Dermanyssina 14
Dermanyssoidea 203
Development 48
Diagnostic characters 56, 62, 64, 66, 68, 69, 70, 71, 73, 74, 87, 92, 95, 96, 99, 105, 110, 114, 116, 118, 119, 120, 127, 130, 134, 136, 141, 147, 150, 152, 154, 155, 156, 157, 158, 163, 164, 165, 166, 167, 171, 180, 186, 189, 190, 191, 192, 193, 203, 204, 207, 211, 212, 213, 214, 215
Dianthus deltoides 136
Diazinon 93, 136, 151
Dichlofluanid 74
Dichlorvos 108, 118, 132, 156
Dicofol 66, 74, 75, 109, 113, 119, 132, 136, 153, 154
Dienochlor 113, 117
Diflubenzuron 151
Diptilomiopidae 127
Disjugal furrow 12, **13**
Dispersal 88, 101, 112, 128, 212, 213
Division of body 12, **13**

East Asia 118
Eggplant 64, 65, 74
Egypt 90, 118, 119, 182, 193
El Salvador 69
Endosulfan 109, 113, 116, 119, 153, 154
Endrin 113, 116, 151
England 90, 96, 164, 181
Entomophagous fungi 58, 64
Entomophthora adjarica 58
Entomophthora thaxteriana 58
Eotetranychus 48, 50, **51**, 52
Eotetranychus lewisi **52**, 69

Eotetranychus sexmaculatus **52**, 70
Epitrimerus alinae 128, 133, **134**
Ereynetal organ **25**
Ereynetidae **25**
Eriophyidae **24**, 127
Eriophyoidea 15, 127
Erythraeidae **25**, 215
Erythraeioidea 215
Ethion 108, 132
Eucharis 115, 147
Euparal 41
Eupathidium 18
Eupodoidea 165
Europe 68, 74, 90, 96, 110, 115, 118, 119, 136, 152, 156, 164, 182, 190, 192, 193, 204, 206
Euseius 172, 173
Euseius citrifolius 95
Euseius concordis 132
Euseius gossipi 135
Euseius hibisci 71
Euseius mesembrinus 71
Euseius ovalis 108
Euseius scutalis 92, **178**, 194
Euseius victoriensis 132
Exochomus pubescens 69

Famulus 18
Fecundity 57, 88
Feltiella 64, 67
Feltiella acarisuga 57
Fenbutatin-oxide 60, 93, 113
Fenpropathrin 154
Fenvalerate 132
Ferns 114
Finland 133, 136
Fixed digit **14**, 212
Flowers 75
Flubenzimine 132
Fluvalinate 108, 132
Formetanate 109
Formothion 108, 132
Fragaria ananassa 118
France 68, 90, 129
Frankliniella occidentalis 187, 189, 208, 213
Freesia 147, 149, 150, 152, 156
French bean 7, 154
French fly 154
Fuchsia 90, 104
Fungicides 74
Fungus gnats 7
Fusarium 147
Fusarium poae 164

Galemdromus annecten 71
Galemdromus porresi 71
Galendromus 172
Galendromus helveolus 71
Galendromus occidentalis 60, 57, 71, **174**, 192
Gardenia 90
Genitalia 16
Gerbera 61, 104, 110, 151, 152, 153, 156, 185
Germany 75, 96, 118, 120, 129
Gladiolus 147, 149, 150
Gloxinia 104, 118
Glyceel 41
Gnathosoma 12, **13**, 47
Gooseberry 120
Grape vine 70
Grapes 5, 71, 184
Grasses 120, 164, 166
Greece 66, 68
Greenhouse whitefly 107
Greenhouses, definition of 3
Guatemala 69

Hallers organ 20
Halotydeus destructor 166, 214
Hawaii 191
Hedera 74, 90, 104, 118
Hedera helix 118
Heinze's P.V.A. 40
Hemitarsonemus 101, **102**

Hemitarsonemus tepidariorum 101, **102**, 114, **115**, 117
Hexythiazox 60, 109
Hibiscus 64, 104
Higher classification of Acari 18
Hippeastrum 115, 116, 156
Hirsutella thompsonii 64
Holothyrida **19**, 20
Homeopronematus anconai 131, 216
Honduras 69
Hong Kong 66, 73
Hoyer's medium 39
Hungary 68, 74, 119, 120, 129
Hyacinth 147, 149
Hypostome 20, 211
Hypoaspis 152, 203, 212
Hypoaspis aculeifer 7, 149, 151, 204, **205**
Hypoaspis miles 7, 204, **206**
Hypoaspis near *aculeifer* 208
Hypoaspis vacua 151, 208
Hysterosoma 12, **13**

Idiosoma 12, **13**, 15, 48
Impatiens 104
India 66, 70, 108, 109, 119
Indonesia 66
Infracapitular setae 20
Infracapitulum 20
Iphiseius 172
Iphiseius degenerans 1, 173, **175**, 180
IPM 59
Iponemus 100
Iran 68, 90
Iraq 68, 70
Ireland 115, 118
Iris 149
Iris kaempferi 101, 120
Israel 68, 90, 129
Italy 118, 119, 120, 129, 184
Ivy 74, 91, 104, 110, 118
Ivy mite 74
Ixodida **19**, 20

Japan 66, 68, 70, 73, 74, 90, 118, 119, 120, 152, 158, 191
Juniperus chinensis 73

Kalanchoë 152, 153
Kanzawa spider mite 66
Karathane 132
Kenya 90
Key to families of Mesostigmata **28**
Key to families of Prostigmata **24**
Key to genera and species of Acaridae **143**
Key to genera and species of Phytoseiidae **174**
Key to genera and species of Tarsonemidae **102**
Key to genera of Tetranychidae **51**
Key to orders of mites **22**
Key to species of *Bryobia* **52**
Key to species of *Eotetranychus* **52**
Key to species of Tenuipalpidae **89**
Key to species of *Tetranychus* **53**
Korea 6, 70, 73, 109, 119
Kuwait 68

Lady beetle 66
Laelapidae **28**, 203
Lambda-cyhalothrin 132
Lasioseius bispinosus 149, 212
Lasioseius fimetorum 212
Lavatera 155, 156
Legs 17, 18
Leptothrips mali 131
Lettuce 6, 158
Lewis mite 69
Lewis spider mite 69
Lactophenol 37
Lactophenol media 40
Libya 69, 90
Life cycle 23, 27, 29, 48, 88, 100, 111, 114, 128, 131, 152, 154, 155, 165, 167, 213, 214, 215, 216
Life history 48, 56, 62, 88, 92, 95, 100,

107, 111, 128, 131, 134, 142, 148, 150, 152, 172, 180, 186, 204, 205, 207
Lily 7, 147, 149, 212
Liuyangmycin 109
Lorryia formosa 166
Lucerne 75
Lycoriella 207, 212

Maceration 38
Macrolophus caliginosus 58
Madeira Islands 69
Malawi 90
Malaysia 66
Mallada basalis 67, 73
Manipueira 109
Mating 57
McGregor strawberry mite 68
Melia azedarach 73
Meloidogyne 215
Meloidogyne javanica 212
Melon 68, 119, 151, 157
Mesonotal scutellum 16
Mesostigmata **16**, **22**, 27, 171, 211
Metaseiulus occidentalis 192
Metastigmata 19
Metasystox 135
Methamidophos 119, 132
Methidathion 149
Methiocarb 109
Methomyl 118, 119
Mexico 68, 69
Microseta 18
Milbex 132
Moldovia 129
Monocrotophos 108
Morocco 68
Movable digit **14**, 211
Mozambique 90
Mushrooms 158
Mycetoglyphus fungivorus **144**, 158

Narcissus 116, 147, 149, 152, 154, 156
Naso 214
Necrosis 50
Neem 66
Neoseiulus 171
Neoseiulus agrestis 108
Neoseiulus barkeri 108, **179**, 189, 153
Neoseiulus californicus 7, 57, 63, 70, 71, 108, 112, 173, **179**, 180, 185, 190
Neoseiulus collegae 71
Neoseiulus cucumeris 7, 73, 112, 136, 153, 171, **179**, 186
Neoseiulus fallacis 57, 66, 67, 73, 112, 132, **179**, 190
Neoseiulus graminis 75
Neoseiulus herbarius 75
Neoseiulus idaeus 92
Neoseiulus longispinosus 57, 63, 66, 67, 108, **178**, 191
Neoseiulus reductus 112
Neoseiulus reticulatus 112
Neoseiulus womersleyi **178**, 191
The Netherlands 68, 73, 90, 96, 115, 118, 193
New Zealand 5, 68, 70, 74, 90, 185, 186, 193, 194
Nicaragua 69
Nicotine 151
North America 5, 70, 110, 118, 193, 204, 206
Notogaster 15
Notostigmata 19

Oceania 70
Oil spray 71, 74
Okinawa Island 66
Oligonychus 50
Oligonychus perditus 73
Oligota flavicornis 67
Oligota kashmirica benefica 67
Oligota oviformis 66
Omethoate 119
Opilioacarida **19**, 20

Opilioacariformes 19
Opiliocarida 19
Opisthonotal shield 16
Opisthosoma 12, **13**
Orchid 147, 152, 158, 180, 185
Oribatida **16**, 19, 20, **21**
Orius 64, 67, 188
Ornithodoros acinus 11
Orthotydeus 166
Oudemans fluid 37
Overwinter 50, 88, 114
Oviposition 50, 57
Oxamyl 113

Pakistan 68
Palp 14
Panama 69
Panonychus 50, **51**, 191
Panonychus citri 71, 213
Panonychus ulmi 71, 193, 216
Papaya 69
Paraphytoseius multidentatus 66
Parasitidae **28**, 211
Parasitiformes 19
Parasitina 14
Parasitinae 211
Parasitoidea 211
Parasitus bituberosus 212
Parasitus fimetorum 149, 212
Parathion 136, 153, 154
Parthenogenesis 88
Peacock mites 167
Peas 166
Penthaleidae 165
Peperomia 113
Pepper 104, 109, 180, 189
Peppermint 92
Pergamasinae 211
Pergamasus 212
Permanent mounts 39
Peru 69
Pest control methods 8
Pest in first method 182
Pesticide-resistant predatory mites 60
Petunia 129
Phalaenopsis 156
Phaseolus vulgaris 65, 107, 181, 183, 212
Philippines 66
Phlox 154
Phoresy 112
Phoretic association 101
Phoretic on insects 212
Phoretic relationship 107
Phoretomorph 164
Phosmet 149
Physogastric females 164
Phytonemus 101, **103**
Phytonemus pallidus 7, 101, **103**, 109, **111**, 119
Phytoptidae 127
Phytoseiidae 15, 21, **28**, 171
Phytoseiinae 171
Phytoseiulus 171, 172, 173
Phytoseiulus macropilis 73, 57, **176**, 194
Phytoseiulus persimilis 7, **18**, 57, 60, 63, 75, 66, 67, 69, 132, **176**, 182, 184, 194
Phytoseius 172
Phytotoxic 61, 73
Pilea 118
Pirimiphos-methyl 113, 117, 149, 153, 156
Plants grown in greenhouses 5
Plastic tunnels 3
Podosoma 12
Poinsettia 69, 206, 207
Poinsettia spider mite 69
Poland 68, 74, 115, 118, 119, 152, 184, 185
Polyphagotarsonemus 99, 101, **102**
Polyphagotarsonemus latus 7, 101, 104, **105**, **106**
Polystichum 114

Populus 185
Portugal 68, 158
Positional mounting 40
Potato 129
Potted plants 6
Precopulation 101
Predatory midges 67
Preparing mites for microscopic study 37
Preserving mites 37
Privet 90
Profenofos 132, 149
Pronematus ubiquitus 131, 216
Pronotal shield 16
Propargite pyridaphenthion 132
Propodosoma 12, **13**
Propylea japonica 73
Prosoma 12
Prostigmata **16**, 19, **22**, 23, 163, 165, 213
Prothiofos 109, 149
Protogamasellus dioscorus 212
Protogamasellus minutus 149
Pseudo-arrhenotokous 172
Pseudotarsonemoidinae 99
Pteris 114
Pteromorpha 16
Pumpkin 64, 151, 157
P.V.A. 40
Pygidial shield 16
Pygmephoridae 163
Pygmephoroidea 163
Pyraclofos 109
Pyrethroids 73
Pyridaben 93, 109, 113, 118, 119

Radish 6
Ramsayellus grandis 21
Raphignathoidea 213
Rearing mites 42
Recinulei 11
Red-legged spider mite 64
Reproduction 92
Resistance 113, 192
Resistance to many pesticides 73
Respiration 16
Rhizoglyphus 7, 141, 142, **143**, 206
Rhizoglyphus echinopus 142, **143**, 149, 208
Rhizoglyphus robini **18**, 142, **143**, 147, 207, 212
Rhododendron 90, 92, 95
Rice 118
Robine bulb mite 147
Root-knot nematode 212, 215
Rose 7, 185, 192
Rove beetle 66
Russia 118, 191
Rutella 20

Saintpaulia 113
Salvia splendens 185
Sarcoptiformes 19
Scarborough lily 115
Schizotetranychus 48
Schradan 113
Sciarid flies 7
Scirtothrips citri 193
Scolothrips 64
Scolothrips indicus 67
Scolothrips sexmaculatus **52**, 58, 69, 131
Scolothrips takahashii 67, 73
Scutellum 15
Scutum 15
Scymnus coccivora 66
Seed oil 73
Sejugal furrow 12, **13**, **26**, 29
Sex determination 112
Sex pheromone 57
Sex ratio 69, 101, 172, 205
Sinningia hybrida 118
Siteroptes 164
Siteroptes avenae **164**
Siteroptidae 163
Sixspotted mite 70

Sixspotted spider mite 70
Skimmia japonica 72
Slide-mounting necessities 34
Solanaceae 129
Solanum tuberosum 74
Solenidion 18
South Africa 68, 69, 90, 115, 166
South America 69, 70, 110, 190
Spain 68, 90, 129
Spermatodactyl **14**, 172
Spermatophores 17, 128, 165
Spermatotreme 14, 211
Spinach 154, 158
Spirothecae 164, 165
Sporidesmium mucosum 157
Sprekelia 115
Sri Lanka 90
Steneotarsonemus 99, 101, **103**
Steneotarsonemus laticeps 101, 115, **117**
Stethorus 58, 64, 73
Stethorus gilvifrons 69
Stethorus pauperculus 66
Stethorus punctillum 58, 73
Stigmaeidae **26**, 92, 213
Stigmata 16, 47
Stratiolaelaps miles 206
Strawberries 7, 65, 68, 109, 110, 113, 180, 193
Strawberry mite 68, 109
Strawberry spider mite 68
Streptomyces griseolus 109
Stylophore **14**
Sulphur 66, 69, 71, 93, 109, 132, 135, 149
Sweden 115
Sweet pepper 109, 184
Sweet peppers 193
Switzerland 68, 74
Symptoms 50, 54, 61, 65

Taiwan 66, 73
Tapinoma melanocephalum 58
Tarsonemidae 99
Tarsoneminae 99
Tarsonemoidea 163
Tarsonemus 99, 101, 103, 119
Tarsonemus bilobatus 101, **103**, 119
Tarsonemus confusus 101, **103**, 118
Tarsonemus floricolus 101, **103**, 120
Tea red spider mite 66
Temporary mounts 38
Tenent hair **18**
Tenuipalpidae **26, 89**
Tenuipalpinae 87
Tenuipalpus 87, 96
Tenuipalpus pacificus **89**, 96
Tetradifon 66, 109
Tetranychidae 15, **26**, 47
Tetranychinae 47, 173
Tetranychoidea 166
Tetranychus 53, 173, 180, 191, 216
Tetranychus atlanticus 68
Tetranychus cinnabarinus 7, 50, **53**, 61, 64, 66, 132, 183
Tetranychus kanzawai **53**, 66, 67, 184, 191
Tetranychus ludeni **53**, 64, 66, 215
Tetranychus turkestani **53**, 68, 214
Tetranychus urticae 7, **18**, **49**, 50, **53**, **55**, 57, 58, 60, 181, 182
Thailand 66, 109
Thelytokous 100, 112, 172
Thiometon 108
Thrips 7, 58, 69, 131, 187, 188, 189
Thrips obscuratus 188
Thrips palmi 188
Thrips tabaci 188
Thumb-claw complex **15**, 23, 47
Tined seta **15**
Tobacco 129
Tobacco whitefly 194
Tomato 7, 61, 104, 118, 128 131, 132, 151, 166, 181, 183
Tomato erineum mite 135
Tomato russet mite 7, 129, 191

Tools for collecting and preparing mites **34**
Trialeurodes vaporariorum 107
Triazophos 109, 113
Trichobothria 18
Triforine 74
Tritosternum **22**, 211
Tropical mite 104
Tuckerella 166
Tuckerellidae 166
Tulip 147, 149 see also *Tulipa*
Tulipa 152, 156
Tunisia 184
Turkey 68, 118
Twospotted spider mite 7, 54
Tydeidae 216
Tydeus 166
Tydeus caudatus 166
Tydeus [=*Orthotydeus*] *californicus* 166
Tydeus praefatus 166
Typhlodromalus limonicus **177**, 193
Typhlodromalus peregrinus 108
Typhlodrominae 171, **174**
Typhlodromips tetranychivorus 66
Typhlodromus 172
Typhlodromus pyri **174**, 193
Tyrophagus 141, 151, 207, 212
Tyrophagus curvipenis 158
Tyrophagus longior **146**, 155
Tyrophagus neiswanderi **146**, 156
Tyrophagus perniciosus **145**, 157
Tyrophagus putrescentiae 145, 152, 155, 189, 213
Tyrophagus similis **146**, 154, 155
Tyrophaus 206

Uganda 90
UK 5, 75, 114, 115, 120, 133, 152, 183, 185, 187, 188
Ukraine 74, 115, 118, 119
USA 5, 68, 69, 70, 72, 73, 90, 114, 115, 118, 120, 136, 156, 182, 185, 188, 190, 192, 193
USSR 157, 182, 187
Uzbekistan 132

Venezuela 90, 107
Verbena 155, 156
Viola 152, 153

Watercress 110
Watermelon 107, 151
Whitefly 58, 189

Xenotarsonemus belemnitoides 101, **102**, 118

Yellow (tea) mite 104
Yugoslavia 68

Zantedeschia aethiopica 185
Zetzellia 213
Zineb 132

inted and bound by CPI Group (UK) Ltd, Croydon, CR0 4YY
16/09/2026
14963431-0001